Cere... ...
biology, simulation and prediction

Best Wish
Tony

Simulation Monographs

Simulation Monographs is a series on
computer simulation in agriculture and
its supporting sience

Cereal aphid populations: biology, simulation and prediction

N.Carter, A.F.G.Dixon & R.Rabbinge

Wageningen
Centre for Agricultural Publishing and Documentation
1982

CIP gegevens

Carter, N. - Cereal aphid populations: biology, simulation and prediction /
N. Carter, A.F.G. Dixon and R. Rabbinge - Wageningen: Pudoc, oktober 1982.
97 p.; 22 cm - (Simulation monographs)
Met lit. opg.
ISBN 90-220-0804-5
(W)
SISO 573.3 UDC 574.9
Trefw.: populatiebiologie.

ISBN 90-220-0804-5

Printed in the Netherlands.

Contents

1 Introduction

1.1 Reasons for studying cereal aphids

The large area of land devoted to cereal crops, with nearly 40% of all arable land under wheat, and the apparent increase in the incidence of cereal pests and diseases over the last twenty years justify a study of cereal pests. In terms of hectarage wheat is one of the most important crops in the world and any loss of yield caused by pests has serious consequences, both locally and world-wide. In Great Britain cereal aphids were not considered as important pests until 1968, when they reached very high levels on wheat (Fletcher & Bardner, 1969). George (1974, 1975) and Kolbe (1969, 1970) have shown that cereal aphids can cause considerable losses of yield in some years. Their abundance, however, varies from year to year (Carter et al., 1980; Rabbinge et al., 1979) and from place to place (George & Gair, 1979). For an effective advisory service knowledge of loss of yield relative to aphid density and the growth of aphid populations on cereals are needed. The latter is dealt with in this study; the former is discussed by Rabbinge et al. (1981). To predict cereal aphid outbreaks it is necessary to understand what causes the spatial and temporal differences in abundance.

Over the last decade considerable changes have occured in the cultivation of wheat in Western Europe. High sowing densities, split nitrogen dressing and top dressing at flowering have resulted in crops that remain suitable for cereal aphids up to the end of crop growth. To prevent losses of yield there has been a marked increase in the amount of biocide (herbicides, fungicides and insecticides) applied to wheat crops. There has also been a tendency to use insurance spraying, i.e. the regular application of biocides at particular crop developmental stages without verifying the presence of pests or disease. This causes an overuse of pesticides, which increases costs and reduces profit per hectare (Rijsdijk et al., 1981) and can result in the development of resistance to pesticides as has been recorded for orchard pests (Helle & Van de Vrie, 1974), and may increase the incidence of secondary pests and diseases(Baronyovits, 1973; Potts, 1977). The improvement in growing conditions has resulted in wheat crops of 9 000-10 000 kg ha^{-1}. This has also had consequences for the development of pests and diseases, as these wheat crops suffer relatively more from pests and diseases than poorer crops (Rabbinge and Rijsdijk, 1982). In the absence of more effective means of controlling the pests and diseases of wheat, pesticides will be used even more frequently in the future. Zadoks et al. (in press) have used a systems approach to predict the incidence of wheat pathogens in the Netherlands (called EPIPRE = EPIdemic PREvention). Those farmers who use this system obtain

1

yields similar to farmers who do not, but with fewer sprays, and hence lower costs.

Throughout Western Europe cereal aphids are serious pests, and in the early seventies studies on their epidemiology were started in several countries. Several simulation models of cereal aphid populations have been developed, for example that of Rabbinge et al. (1979), which was developed for use in conjunction with EPIPRE in the Netherlands. The model presented in this book was developed for Norfolk, England.

After Lincolnshire, Norfolk is the most important cereal growing county in England and Wales (Anonymous, 1980). Although more barley is grown than wheat, there were still over 85 000 ha under wheat in Norfolk in 1979. This was a major factor in the decision to work on cereal aphids at Norwich. From 1976 to 1980, with the notable exception of 1979, when *Metopolophium dirhodum* was the most common aphid, the English grain aphid, *Sitobion avenae,* has been the most numerous cereal aphid species on wheat in Western Europe, and as a consequence most research has been done on this species.

1.2 Aims of the study

Warning schemes are based on three components: monitoring, forecasting and communication. Monitoring involves sampling the pest either outside the crop (at overwintering sites or while they are dispersing) or on the crop. Monitoring can be carried out by the farmer or extension worker or, especially in the developmental phase, by the research scientist. To be successful the monitoring must be easy to carry out, reliable, quick and cheap (Rabbinge, 1981). The time spent monitoring, the detail and the frequency depend on the value of the crop and the time from sowing to harvest. Approximately 1 h field^{-1} yr^{-1} (field $\leq$ 20 ha) of monitoring cereal aphids is acceptable. A rapid flow of information between farmers, agricultural advisers and researchers is vital. If there is a long delay between monitoring and issuing advice the scheme will not be attractive to farmers (Welch & Croft, 1979).

Our aims in this monograph are to explain the population development of *S. avenae* on cereals and to indicate how cereal aphid outbreaks might be predicted. To achieve this comprehensive models are developed which may be reduced to simple decision rules and incorporated into a decision-making process involving the whole system. Way & Cammel (1974) have remarked on the problems of early forecasting of outbreaks of *S. avenae,* due mainly to the widespread distribution of its overwintering hosts and the lack of reliable long-term weather predictions (Lamb, 1973).

Several attempts have been made to predict outbreaks; these have been reviewed by Dixon (1977) and Carter & Dewar (1981). Dean (1973b) found no relationship between the numbers of cereal aphids caught in suction traps in May and June and the peak densities achieved on nearby crops. However suction trap catches, do indicate the start of immigration into crops in spring. This is correlated with the weather conditions prevailing in the early part of the year (Sparrow, 1974; Walters, 1982). As the number of suction traps increases, especially in Germany, France and the Netherlands, an overall picture of cereal aphid migration will develop that will result in more accurate predictions of the timing of the immigration and in the future more reliable estimates of the numbers flying. However more information on the migratory behaviour of ceral aphids is needed to predict the time of arrival and number of aphids that will colonize specific fields.

At present, cereal aphid population forecasting in Europe uses a critical point model, i.e. the number of aphids at flowering. Five or more aphids per ear and increasing at that stage is thought to indicate that the aphids will become abundant enough to justify spraying (George, 1974, 1975). The evidence for this scheme is rather tentative, with no consensus on the critical infestation level at flowering, e.g. in Belgium and France it is 15-20 aphids per tiller, in the Netherlands it is 70% of tillers infested (5 aphids per tiller) (Rabbinge & Mantel, 1981). Most field and laboratory observations indicate that flowering in cereals is a critical period for *S. avenae*, but there is no consistent relationship between aphid numbers at flowering and the peak aphid population (see Chapter 5).

An alternative, though partly complementary, approach is to use sensitivity analysis in combination with a simulation model to study aspects of the system that are poorly understood. Data on the biology of *S. avenae* and its important natural enemies have been combined in just such models (Carter, 1978; Rabbinge et al., 1979). The predictions of these models are validated by comparison with trends in cereal aphid numbers collected over a number of years from different localities. The importance of each modelled component of the aphid's biology is evaluated by varying the value assigned to that component.

Such a sensitivity analysis may eventually result in cumbersome simulation models being replaced by simple decision rules. This hierarchical approach, i.e. comprehensive explanatory models → summary models with many descriptive elements → decision rules (Loomis, et al., 1979), will lead to forecasts as reliable as the EPIPRE system (Subsection 5.2.2.).

Further field and laboratory studies are needed on morph determination, natural enemies and the interaction of biotic and abiotic factors on the loss of yield. In Chapter 2 the biology of the cereal aphid ecosystem is described in some detail, but for a fuller account see Vickerman & Wratten (1979) and Carter et al. (1980). A simulation model for the development of *S. avenae* populations on winter wheat, with reference to the other aphid species where relevant, is presented in Chapter 3. The predictions of the model, their validation and sensitivity to changes in parameter values are considered in Chapter 4. In Chapter 5 the findings of the simulation study are discussed, some of the simple decision rules that have been developed for *S. avenae* are introduced, examples of their application are given and areas for future research are indicated. A listing of the simulation model is given in Appendix A.

2 Biology of the cereal aphid system

2.1 Aphids

Three species of cereal aphid commonly infest cereal crops in Western Europe: the English grain aphid *(Sitobion avenae)*, the rose grain aphid *(Metopolophium dirhodum)* and the bird cherry-oat aphid *(Rhopalosiphum padi)*. The numbers of alate adults of these three species caught from the beginning of May to the end of July by white (1942-1946) and yellow (1951-1969) sticky traps (Heathcote, 1970) and a suction trap at Rothamsted (1968-1971) were used by Dean (1974a) to determine annual trends. He found that these results did not always correspond to crop infestation levels as reported by the British Ministry of Agriculture, Fisheries and Food. However, the suction and sticky trap catches give many years of quantitative information on the timing and abundance of cereal aphids.

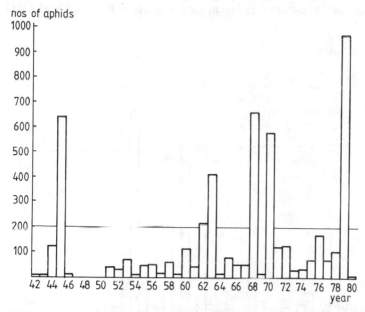

Figure 1. Number of alates of *M. dirhodum* caught each year from 1942-1980 at Rothamsted. From 1942-1971 the numbers presented are sticky trap catches; from 1972 onwards the equivalent catches predicted from suction trap catches using the relationship between suction and sticky trap catches for the period 1965-1971, when both types of trap were used. Horizontal line represents an arbitrary level for assessing outbreaks.

As both sticky and suction traps were used at Brooms Barn and Rothamsted (Fig. 36) from 1965-1971 and the degree of association between the catches taken by the two types of trap is good (Brooms Barn: *M. dirhodum* $r = 0.7$, *S. avenae* $r = 0.9$; Rothamsted: *M. dirhodum* $r = 0.8$, *S. avenae* $r = 0.9$), it is possible to standardize the catches and obtain an indication of changes in abundance of *M. dirhodum* and *S. avenae* from year to year from 1942 to 1980 (Figures 1 and 2). There is a weak correlation between the year to year abundance of *M. dirhodum* and *S. avenae* ($r = 0.42$, $n = 35$, $p < 0.02$), with years when only one species is common. Of the three species, *S. avenae* is considered to be the most common pest of wheat. *M. dirhodum,* however, is more abundant than *S. avenae* in trap catches. Their relative abundance possibly reflects the proportion of the cereal hectarage that is under barley (Figure 3), the preferred cereal host of *M. dirhodum* (Dean, 1973a). Since 1942 the area planted with cereals and the proportions made up of barley, oats and wheat have changed (Figure 4). These changes are likely to affect the abundance of the two species of aphid; the decrease in the area sown to barley adversely affecting *M. dirhodum.*

Figure 2 shows that *S. avenae* was more abundant in the late 1960s than previously, even taking the increase in cereal hectarage into account. The numbers of *M. dirhodum* have also increased since 1968. Although it is unlikely that the trap catches accurately reflected aphid numbers on the crop, they can be used to indicate the frequency of outbreaks. The arbitrary levels in Figures 1 and 2 and

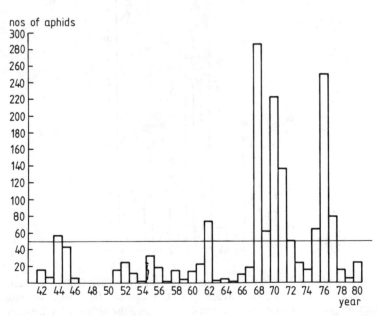

Figure 2. Number of alate *S. avenae* caught each year from 1942-1980 at Rothamsted. Details of traps are described in legend to Figure 1.

6

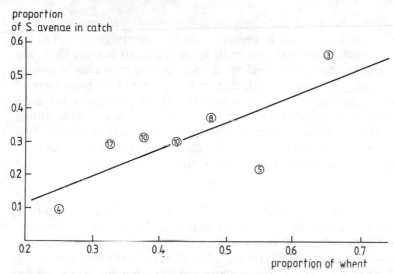

Figure 3. Relationship between the proportion of *S. avenae* and *M. dirhodum* in the suction trap catches at Brooms Barn and Rothamsted and the proportion of wheat to barley each year in the county in which the traps are situated. The 52 place years are grouped into 7 classes according to the proportion of wheat; 4, for example, is the number of observations.

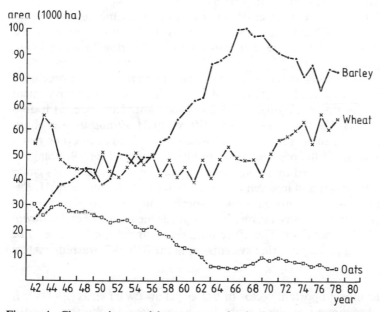

Figure 4. Changes in cereal hectarage under barley, oats and wheat in Hertfordshire from 1942-1978.

7

a correction for the hectarage under cereals, indicate that there have been more outbreaks of *M. dirhodum* and *S. avenae* in the last twenty years than in the previous twenty years. Thus both species of aphid appear to be more abundant and more frequently achieve high levels of abundance than previously — at least in the trap catches. This may be partly due to the increase in the proportion of the land under cereals (currently 46% in Suffolk). The main reasons, however, are changes in agricultural practices, which have resulted in more aphids settling and an increase in their developmental and reproductive rates on cereals. This view is supported by an increase in cereal aphid outbreaks in those parts of Europe where there has been a decrease in the hectarage under cereals, e.g., the Netherlands.

M. dirhodum was the most abundant cereal aphid in Western Europe in 1979 (Figure 1) and is in some other parts of the world the commonest species every year, e.g. in South America (Zuniga & Suzuki, 1976). It overwinters either viviparously on cereals and grasses (Dean, 1978) or as eggs on roses, its primary host (Hille Ris Lambers, 1947). Egg-hatch occurs in March and April and the first generation is totally apterous. The proportion of alate adults in the second generation is variable but the third generation adults are all alate. Migration to Gramineae occurs in late spring and early summer. In autumn, gynoparae and males appear in response to short day-length and low temperatures (Elkhider, 1979). These morphs return to the primary host where the gynoparae produce oviparae, which lay eggs after mating with the males.

M. dirhodum feeds on the leaves of cereals and grasses, moving on to new leaves as they appear (Dedryver, 1978), in contrast to *S. avenae,* which feeds predominantly on the ears. *M. dirhodum* infestations of spring barley and winter wheat can cause significant losses of yield (George, 1974; Wratten, 1975). Because in most years in Britain this aphid is rare on wheat, it was ignored until the outbreak of 1979. In the Netherlands this species is more abundant on wheat than in Britain, but was not thought to cause significant yield losses as it does not feed on the ears (Vereijken, 1979). In 1979, when *M. dirhodum* was abundant and *S. avenae* virtually absent, yield losses were as high as in other years, so a detailed analysis of the previous years results was undertaken (Rabbinge & Mantel, 1981). This revealed that both *M. dirhodum* and *S. avenae* may cause considerable yield losses and in seven out of 21 fields in different years *M. dirhodum* did most damage. Thus both aphid species contribute to yield loss. In 1979 in Great Britain there was a severe loss of yield, the numbers of *M. dirhodum* reaching 50-300 per tiller. Therefore both cereal aphid species have to be included in warning and monitoring systems, as in the EPIPRE warning system in 1980.

R. padi is the most important vector of barley yellow dwarf virus (BYDV). It is one of the commonest aphids caught in suction traps in Western Europe (Dean, 1974a) and is potentially a very serious cereal pest (Dean, 1973a; Leather

& Dixon, 1981). However, it only achieves high numbers on cereals in Scandinavia (Markkula, 1978). It can either overwinter as an egg on bird cherry *(Prunus padus)*, its primary host, or viviparously on cereals and grasses (George, 1974). The egg mortality is 70%, occurring at a rate of 1.8% per week during a winter of 20 weeks, rising to 3% in spring (Leather, 1980). Similar high egg mortality has been recorded for other aphids (Dunn & Wright, 1955; Way & Banks, 1964). The eggs hatch in March and April, and emigrant aphids are induced by crowding and poor nutrition (Dixon, 1971; Dixon & Glen, 1971). Few of these alatae colonize cereals in the U.K. Its nonpest status in Western Europe is puzzling as it thrives on cereals in the laboratory. However, it has a greater preference for grasses (Leather & Dixon, 1981). In autumn *R. padi* like *M. dirhodum* produces gynoparae and males in response to short day-length and low temperature (Dixon & Glen, 1971; Dixon & Dewar, 1974). These morphs return to bird cherry, where the gynoparae produce oviparae, which mate with the males and then lay eggs.

S. avenae is monoecious on Gramineae, on which it may overwinter as viviparae or as eggs (Figure 5), but no information is available on their relative im

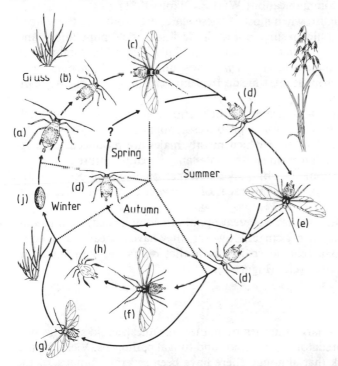

Figure 5. Life cycle of the English grain aphid, *S. avenae*. a: fundatrix. b: apterous fundatrigeniae. c: emigrant. d: apterous exule. e: alate exule. f: gynoparae. g: male. h: oviparae. i: egg.

portance (Dean, 1974a). It is very difficult to find either aphids or eggs on grasses or cereals in Norfolk during the winter. As a consequence, it has not been possible to monitor spring populations to predict possible levels of infestation of cereals. However, this might be possible in those parts of Europe where overwintering aphids have been found on grasses, e.g., in Belgium and the Netherlands.

In most years, from the end of May until the end of June, alates of *S. avenae* colonize winter wheat, in preference to most other cereals (Carter, 1978). Alate aphids are caught in suction traps, usually before they are found in cereal fields (George, 1974). At the start of immigration, wheat has not usually headed and the alates settle on the leaves. As the ears emerge they are colonized by the alates and their nymphs. Most of these nymphs develop into apterous adults whose reproductive rate is higher than that of alate adults (Wratten, 1977). The reproductive rate of *S. avenae* is higher on the young ears (up to the start of the milky-ripe stage) of cereals than on the leaves or the older ears (Vereijken, 1979; Watt, 1979a). Reproductive rate also decreases with increase in aphid density (Vereijken, 1979). As the aphid density rises and the crop ripens an increasing proportion of the nymphs born to these apterae develop into alate adults (Ankersmit, personal communication; Watt & Dixon, 1981).

In July and early August, when most of these alates are produced, the winged aphids leave the crop, which results in a rapid decline in field populations and large catches of cereal aphids in the suction traps. This emigration is induced by a combination of high aphid density and poor host quality. The large number of natural enemies in the crops at this time destroy any remaining aphids (McLean, 1980).

The autumn migration, as indicated by the suction trap catches, is very small. Under short day conditions *S. avenae* produces gynoparae first, followed by a short break in reproduction, after which mostly males are produced. This sequence of morph production ensures that males and oviparae mature at approximately the same time (Watt, 1979b). Ankersmit (personal communication) and Watt (1979b) have evidence that some races of *S. avenae* do not produce sexual forms.

The life cycles of these species of cereal aphid are complicated, each having eight morphs, including the egg stage. However, they have a great deal in common; figure 6 presents a generalized life cycle that omits the daunting complexity of the detailed life cycles (Figure 5).

2.2 Natural enemies

Cereal aphids have a large number of natural enemies: aphid-specific predators, polyphagous predators, parasitoids and fungal pathogens. Vickerman & Wratten (1979) remark that although there have been several estimates of the numbers of natural enemies, little has been done to quantify their effects on cereal aphid population growth. As with the aphids, little attention has been

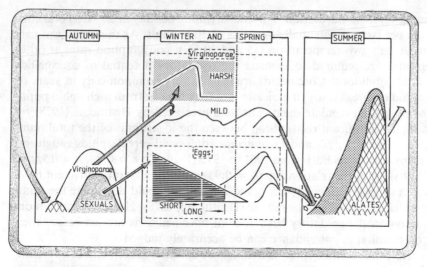

Figure 6. Generalized cereal aphid life cycle.

given to natural enemies in habitats away from cereals, especially in winter and spring. Vickerman & Wratten (1979) suggest that natural enemies might exert a major interseason effect on aphid abundance but their contribution to aphid population regulation does not seem promising at present. Further combined field and laboratory studies are needed to elucidate the potential of these organisms.

2.2.1 Aphid-specific predators

Much of the work carried out on the natural enemies of cereal aphids has centred on this group. They belong mainly to three families: Coccinellidae (Coleoptera), Syrphidae (Diptera) and Chrysopidae (Neuroptera). The larvae and adults of the first family, but only the larvae of the latter two, eat aphids. There have been no detailed studies of their population dynamics. These predators are also subject to mortality caused by other predators, parasitoids and pathogens; these in turn have been studied even less.

Coccinellids arrive in cereal fields in spring, about the same time as the aphids, but in cool summers they may take several weeks to become reproductively mature. The threshold density of aphids, below which coccinellids will emigrate, is not known. Fecundity, however, is directly related to aphid consumption (Dixon, 1959). Although they sometimes lay their eggs in conspicuous places it is difficult to get a reliable estimate of egg density. McLean (1980) has studied the consumption rates of the different instars of *Coccinella 7-punctata* and *C. 11-punctata* in the laboratory when supplied with excess aphids. The fourth larval (the last) instar is the most voracious, with that of *C. 7-punctata*

11

consuming up to 40 third instar *S. avenae* nymphs per day at 20 °C. Total consumption per larva is potentially in excess of 300 aphids. As this study was carried out at only two temperatures (15 and 20 °C), consumption rates at other temperatures are required to evaluate the regulatory potential of coccinellids under field conditions. Coccinellids are usually very common only in years of cereal aphid outbreaks, which indicates that they benefit from high aphid populations rather than regulate them. The data presented by Heathcote (1978) reveals a highly significant relationship between the logarithms of the total number of *Coccinella, 7-, 10-* and *11-punctata* adults and cereal aphids caught on sticky traps at Brooms Barn from 1961 to 1975 (Figure 7: $r = 0.74$, $n = 15$, $p < 0.01$). The slope of the relationship ($b = 0.93$) is not significantly different from 1.0, which supports the view that aphid abundance could determine the numbers of coccinellids and not the reverse (Heathcote, 1978). However, much more information on the predatory behaviour of coccinellids is needed before their effect on cereal aphid abundance can be accurately judged.

The percentage of time spent searching, the speed of search and the developmental rate of the coccinellids are temperature dependent, which means coccinellids are likely to be more effective in warm sunny years than in cool cloudy ones (McLean, 1980). However, as aphid development and reproduction are also temperature dependent, this view might not be correct, and indeed Vickerman

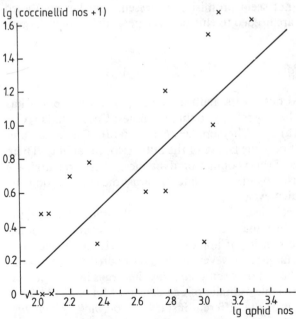

Figure 7. Relationship between the numbers of coccinellids and cereal aphids caught on sticky traps at Brooms Barn, 1961-1975 (after Heathcote, 1978).

12

& Wratten (1979) suggest that natural enemies are more important in years with cool summers. A simulation model of the interaction between coccinellids and aphids is one way of testing which of these two ideas is correct. But this must wait until more information on the natural enemies is available.

Only the larvae of syrphids eat aphids. The adults feed on nectar (energy source) and pollen (protein source). *Episyrphus balteatus* is the most important species in cereal fields in Western Europe, although other species are commonly found, e.g. *Melanostoma mellinum, Sphaerophoria scripta* and *Metasyrphus luniger*. Adult female syrphids will oviposit on plants contaminated with honeydew (Schneider, 1969), but there is little information on either their fecundity or the development and consumption rates of their larvae in cereal fields. In addition syrphid eggs and larvae are difficult to find and it is usual for more pupae to be found than would be expected from the numbers of larvae recorded.

Chrysopa carnea is found in cereal fields, but only at low densities. Only the larvae are aphidophagous, but they are very mobile and consume aphids very rapidly. Even less is known of their biology than for syrphids, but as they are rare they are unlikely to have a significant effect on cereal aphid populations.

2.2.2 Polyphagous predators

This group of predators includes Carabidae and Staphylinidae (Coleoptera), Dermaptera, Araneae and Acari. There is an inverse relationship between arthropod diversity (positively correlated with the percentage of predatory individuals) and the density of apterous aphids in June (Potts & Vickerman, 1974). Although many of these predatory species consume aphids it is not known how many each consume (Sunderland, 1975). Exclusion of most ground predators has often resulted in higher aphid densities, but usually only when aphid densities were low. In 1979, however, at Rothamsted, the presence of ground predators reduced the aphid peak population of *M. dirhodum* from more than 200 to 50 per tiller. Rather surprisingly though, there was no difference between the treatments at North Farm in West Sussex (Sunderland et al., 1980). Except in the early part of the season, ground predators had no effect on aphid population build-up in Norfolk in 1977. This was because the large number of immigrant cereal aphids swamped any effect the predators had (McLean, 1980).

There is little relevant information on the searching behaviour or consumption rates of these predators and, as McLean (1980) remarks, few field studies take them into account. Polyphagous predators are also easily overlooked in field surveys as many of them are more active at night (Vickerman & Sunderland, 1975). As a large number of species are involved, the evaluation of their effect on aphid populations is difficult and this will remain the case until more detailed field and laboratory studies are carried out on these species.

2.2.3 Parasitoids

The primary parasitoids of cereal aphids belong to two families of Hymenoptera: the Aphelinidae and the Aphidiidae. The latter is the more important in Europe (Carter et al., 1980), but, as with the other natural enemies of cereal aphids, little is known of their biology, and their taxonomy is confused (Powell, personal communication). The species composition of parasitoids changes from year to year and from place to place (McLean, 1980), but *Aphidius ervi, A. picipes, A. uzbekistanicus* and *A. rhopalosiphi* are the most common, although *Praon volucre* may also be quite numerous. There is very little quantitative information on them as parasitoids of cereal aphids.

A. rhopalosiphi is short-lived, with a maximum adult life-span of twenty days at 20 °C. It is very fecund, realizes a fecundity of more than 200 eggs, and prefers to oviposit in young aphids, i.e., those in the first three instars (Shirota et al., to be published). Until similar studies have been carried out on several species of parasitoids, at a number of temperatures, it will be difficult to evaluate their effect on cereal aphid populations. Once more is known about the biology of these species of parasitoids, it may be possible to explain the changes in species composition that have been observed from year to year.

One major factor limiting the increase in the numbers of parasitoids is the action of hyperparasitoids, which can rapidly increase in numbers and destroy many parasitoids at the end of the season. However, the degree of hyperparasitism varies from year to year (McLean, 1980) and so it is difficult to predict the effect of hyperparasitoids early in the season. Jones (1972) suggests that the aphid-parasitoid-hyperparasitoid interaction is very important in determining whether an aphid outbreak will occur. *Dendrocerus carpenteri* is probably the most common hyperparasitoid but a number of other species can also be numerous.

2.2.4 Fungal pathogens

These belong to the genus *Entomophthora*. Quite frequently the first signs of disease occur late in the season (Dean & Wilding, 1971, 1973), although Latteur (1976) suggested that occasionally *Entomophthora* have a significant effect on aphid population build-up. Fungal disease seems to be most important in years with high rainfall during the summer. The introduction of *Entomophthora* into cereal crops, early in the season has not met with much success (Dean et al., 1980), but their use in biological control deserves more attention. The use of other biological control agents could prove costly but the resting spores of *Entomophthora* and other fungi are relatively easy to propagate, store and apply.

2.3 The crop

There are several models of crop growth, but few of crop development. The

14

latter are descriptive formulae including, in some cases, field data. Maas and Arkin (1980) developed a general model of wheat development, whereas Seligman & van Keulen (1981) considered the effect of water shortage and nitrogen balance on the rate of development of winter wheat. The latter is suitable for use with an aphid model but there is insufficient information on the availability of nitrogen and water in the field. Therefore, a simple algorithm is used in this study. It is assumed that crop growth is not a limiting factor for the build-up of aphid populations. The crop developmental stage, however, is known to affect aphid biology, especially wing induction, fecundity, survival and adult longevity (Ankersmit, to be published; Vereijken, 1979; Watt, 1979; Watt & Dixon, 1981). Thus crop developmental stage, described by the decimal code of Zadoks et al., (1974), is a measure of the quality of this food source for cereal aphids.

3 The model

3.1 Introduction

Simulation models or dynamic models (Jeffers, 1978) have only recently been used in ecological studies, as they are dependent on computers to solve large numbers of equations quickly. Models are used in systems analysis to study the interactions between state variables, which quantify all properties that describe the state of the system. The underlying assumption is that the state of an ecosystem at any particular time can be expressed quantitatively and that changes in the system can be described in mathematical terms (de Wit & Goudriaan, 1978). These changes in the system, are controlled by rate variables, which are assumed to remain constant over small time intervals. Therefore the time interval used is dependent on the rates of change of the system, but for practical reasons it must not be too small. The duration of the time step is dictated by the rate of change calculated from the time coefficient (de Wit & Goudriaan, 1978). This is difficult to determine for large systems, such as *S. avenae* populations, and in such cases trial and error is used to determine the most appropriate time interval, which in this case was found to be 1 hour.

Besides state and rate variables, there are auxiliary, output and driving variables. Driving, or forcing, variables are those that are not affected by processes within the system but characterize the influence from outside. These may be for instance the temperature or the temperature sum. Auxiliary variables are those which are used to simplify processes to increase understanding. Output variables are those quantities which the model produces for the user and can be any of the aforementioned variables.

Simulation models may either be deterministic (dependent on proportions) or stochastic (dependent on probabilities) or a mixture of both where some processes are treated stochastically. In most cases it is only the mean value that is of interest and not the range of possible values. When one wants to know the range of possible values a stochastic simulation is needed. Also, when some of the relationships between the variables with a stochastic character and state variables are curvilinear deterministic models give erroneous results (Fransz, 1974; Rabbinge, 1976). In such cases stochastic models are needed. However stochastic models use a lot of computer time as each simulation has to be repeated many times (up to 1000) to produce the range of responses. To overcome this other methods have been developed, which are basically deterministic, but make use of classes of individuals (Rabbinge & Sabelis, 1980).

16

We use a deterministic model as the relations between rates and the forcing functions incorporated in the model are linear. When a more detailed predator-prey relationship is available the model will have to be modified. Barlow & Dixon (1980) in their simulation model of lime aphid populations also used a deterministic approach, with discrete time steps, as in our present model; a constraint imposed by the use of the computer language FORTRAN IV. At the University of East Anglia (UEA) it is difficult to run large models written in simulation languages, such as CSMP, which also treat processes in a discrete way but may make use of more complicated numerical integration procedures.

The other *S. avenae* population model (Rabbinge et al., 1979) has the same structure but is written in CSMP. It incorporates a range of experimentally determined developmental times (dispersion in time) (Goudriaan, 1973) and can be run on all computers with a FORTRAN compiler. As dispersion in time had little effect on predictions it is omitted from the present model (Carter & Rabbinge, 1980).

The model describes the population growth of *S. avenae* on winter wheat in summer. No attempt has been made to simulate population growth on grasses or to model the relationship between years. The model starts with the colonization of the crop by alates in May and June, together with allowances for any aphids which have overwintered on the crop.

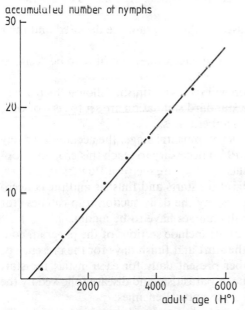

Figure 8. Cumulative daily total production of nymphs by *S. avenae* reared on wheat seedlings in a glasshouse at 20.0 ± 5.0 °C.

17

There are two ways of initializing the aphid population.
- The number of colonizing alates can be estimated from the catches of the nearest suction trap.
- The initial numbers of aphids can be derived from field counts as long as subsequent immigration is measured.
The former is used by Carter et al. (1980) and the latter by Rabbinge et al. (1979).

The numbers of aphids in each instar are stored in elements in one-dimensional arrays, in addition to an associated one-dimensional array containing their ages in hour degrees (H°). At each iteration, the numbers in each element are moved up one in the array after correcting for developmental mortality. The corresponding elements in the one-dimensional aphid-age arrays are also updated. The survival rates of the nymphs and adults are temperature dependent and are also affected by the crop developmental stage: as the crop ripens the survival rates decline. Reproduction is also crop- and temperature-dependent, but it does not depend on adult age as it is assumed that they do not live long enough for this to become important (Figure 8; Wratten, 1977).

As the crop ripens and the aphid density increases the proportion of nymphs that develops into alates increases. It is assumed that these alates emigrate before depositing nymphs. The effects of coccinellids, parasitoids and disease are modelled, albeit in a simple way.

The model, a listing of which is given in Appendix A, consists of a series of steps:
1. Model initialization. The arrays used in the program are declared and their elements are zeroed.
2. Data input. These are variable and parameter values that have to be declared at the start of each simulation:
   - the days on which the simulation will start and finish, followed by the sensitivity analysis factors, which in a standard simulation are set to zero or one;
   - daily maximum and minimum temperatures;
   - latitude of the site, initial crop developmental stage, the accumulated day degrees above 6 °C (Anonymous, 1943) necessary to reach this crop developmental stage and the number of tillers per square metre at flowering or after;
   - the days on which aphid immigration starts and finishes and the concentration factor (normally 40), followed by the daily suction trap catches (for initialization with field counts small changes have to be made).
   - three parameters are used to omit or include sections of the program concerned with the natural enemies: the start and finish days for the presence of coccinellids followed by the number present daily for each instar; the start and finish days of the mortality due to parasitism and disease followed by the hourly mortalities attributable to these natural enemies;
   - the times of sunrise for each day are calculated and stored in a one-dimensional array. ·

18

3. Hourly temperatures. A sine curve fitted through the minimum (at sunrise) and maximum (at 1400 h) temperatures. The number of H° for a particular hour is the temperature plus 3.6.

4. Immigration. The number of immigrant aphids landing per tiller each day is calculated using the number of S. avenae caught in the nearest suction trap, or, when the second method of initialization is used, from the number of alates counted in the crop.

5. Development and survival. This involves hourly changes in the numbers and ages of aphids in each age class, dependent on the temperature and crop developmental stage. The number of aphids entering the adult instar (both apterous and alate) is reduced due to the mortality caused by parasitism and disease. Alate adults emigrate immediately after they become adult.

6. Reproduction and morph determination. The numbers of new nymphs produced each hour are dependent on temperature, crop developmental stage and adult morph. The alate and apterous adults are treated separately but their nymphs are combined. The proportion that develop into alate adults is determined by a multiple regression equation that incorporates aphid density and crop developmental stage.

7. Predators. Mortality due to the activity of aphid-specific predators is determined using a subroutine (listed at the end of the main program).

8. Output. This consists of the numbers and morph of aphids in each instar together with crop developmental stage, and the effects of predators, parasitoids and disease. It is given for 1200 h each day for comparison with field results.

9. Crop development model. At the end of each day the developmental stage of the crop is updated dependent on the accumulated day degrees (D°) above 6 °C up to and including that day.

10. Input variables. At the end of a simulation, numbers of aphids immigrating, predator numbers, mortality due to parasitism and diseases and daily minimum and maximum temperatures are also printed.

3.2 The aphid

3.2.1 Immigration

The submodel

For initialization the procedure using suction trap catches is given; the method using actual field counts is described by Rabbinge et al. (1979).

The timing and the daily number of alates colonizing a cereal crop are estimated from the number of aphids caught during the immigration period in a suction trap, 12.2 m above the ground. The number of alates colonizing one million tillers (ALATIM) each day (the number of tillers per square metre is needed here) is calculated by multiplying the daily suction trap catch by the deposition factor (TAYPAL) and the concentration factor. ALTIM is the corresponding number of alates per tiller. It is assumed that these alates have recently moulted and that after colonizing the crop they do not emigrate but stay until they die.

The data

Aphids fly at heights of up to more than 1000 m. Their mean density diminishes with height according to the equation (Taylor & Palmer, 1972)

$$F(z) = C(z + z_e)^{-\alpha}$$

where $F(z)$ is the density at height z, C a constant related to population size, α a density gradient and z_e a measure of the departure from linearity on logarithmic co-ordinates. Thus the number of aphids flying over a given area of ground can be obtained by integration. In those species of aphid in which the density-height profile has been measured the mean daily density gradient (α) ranges from -0.5 to -1.5. In the absence of density-height profiles aphid numbers can be estimated from the equation

$$\lg D = a + b \lg Z$$

in which b has the value -1 and D is aphid density and Z height. The rate at which the aphids are deposited on the ground depends on their flight time. Deposition rates (in aphids ha^{-1}) for several mean flight times (0.5 to 24 h) and density gradients (0 to -2.0) have been calculated (Taylor & Palmer, 1972) and are presented in Table 1.

To calculate the deposition rate it is assumed that the flight time is 2 h, the average flight time for aphids in southern England (Taylor & Palmer, 1972), and the density-height profile is -1.0, which gives 237 alates ha^{-1} (96 acre^{-1}) for every aphid caught in a suction trap. As the number of tillers per square

Table 1. Number of aphids landing (ha^{-1}) equivalent to one aphid caught in a 12.2 m suction trap.

Density gradient (b)	Mean flight time (h)						
	0.5	1.0	2.0	4.0	8.0	12.0	24.0
0	10 315	5157	2579	1289	645	430	215
-0.5	1660	830	415	207	104	69	35
-1.0	948	474	237	119	59	40	20
-1.5	2016	1008	504	252	126	84	42
-2.0	10 315	5157	2579	1289	645	430	215

After Taylor & Palmer (1972).

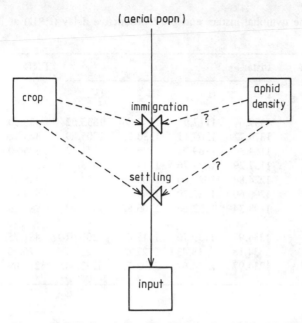

(aerial popn)

crop

immigration

aphid
density

?

settling

?

input

Figure 9. Relational diagram for cereal aphid immigration.

metre is known for each field the number of alates per million tillers can be calculated – this is the deposition factor. This, however, underestimates the number of alates landing on cereals by a factor of about 1/40. The predicted number
of alates is thus multiplied by 40 (the concentration factor), which has been
constant for different wheat varieties from 1976 to 1980. Figure 9 is the relational diagram for immigration.

3.2.2 Development and survival

The submodel

Development in insects is primarily determined by temperature, so that the
stage of development can be calculated by using a temperature summation technique, usually accumulated day or hour degrees (above a threshold temperature). Not all individuals of a species develop at the same rate. With *S. avenae*
the variance of the mean rate is small (Dean, 1974b; Rabbinge et al., 1979) and
in a sensitivity analysis it was shown to have little effect on predicted population
trends (Carter & Rabbinge, 1980). Therefore only mean values were used. De
Wit & Goudriaan (1978) describe this as a BOXCAR routine with no dispersion.
With this approach, aphids of the same physiological age remain together and
are moved from one BOXCAR, or array element, to the next at each iteration.

21

Table 2. Duration (H°) of the nymphal instars and pre-reproductive delay (PRD) at various temperatures.

Temperature (°C)	H° h⁻¹	Instar				PRD
		I	II	III	IV	
10.0	13.6	1339.60	1117.92	1249.84	1335.52	435.20
12.5	16.1	1379.77	1209.11	1130.22	1205.89	325.22
15.0	18.6	1164.36	1169.94	1076.94	1231.32	306.90
17.5	21.1	1137.29	1088.76	1114.08	1375.72	381.91
20.0	23.6	1224.84	1073.80	1005.36	1274.40	448.40
22.5	26.1	1200.60	1145.79	1143.18	1297.17	579.42
25.0	28.6	1198.34	1172.60	1106.82	1384.24	985.18
Mean		1234.97	1139.70	1118.06	1300.61	481.75
Standard error		34.18	18.43	27.89	25.97	76.96
Accumulated duration		1234.97	2374.67	3492.73	4793.34	5275.09

After Dean (1974b)

In the *S. avenae* model, development times, in H°, are constant for each nymphal instar (Table 2), but the longevity of the adult depends on the crop developmental stage, and becomes very short towards the end of the season.

Survival is calculated hourly from the number of H° for that particular hour (Figure 10). For computing convenience updating of the population arrays starts with the oldest age class of the adult instar and works backwards to the

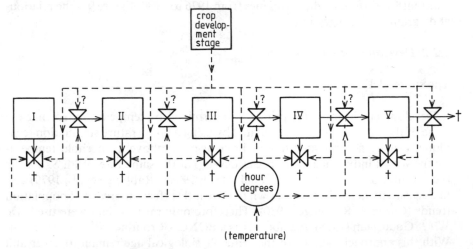

Figure 10. Relational diagram for development and survival of cereal aphids.

youngest age class of the first instar. Mortality is taken into account during the updating procedure:

> $NUMBERS\ (I)\ =\ NUMBERS\ (I\ -\ 1)\ \times\ PROPORTION\ SURVIVING$
> $AGE\ (I)\ =\ AGE\ (I\ -\ 1)\ +\ H°.$

The age is checked against the limit for that particular instar. If the age is greater than this limit the aphids in that age class are moved into the first age class of the next instar, their age is set to zero and the original array elements are zeroed. If these aphids are adults then they are removed. This process results in the first class of the first instar nymphs becoming vacant, ready for the production of new nymphs. All fourth instar alatiform nymphs emigrate on moulting to the adult stage thus the only adult alates recorded in the simulation are those which migrated into the crop.

The data

The relationship between developmental rate and temperature was calculated for cereal aphids reared on leaf discs of barley (cv. Proctor) (Dean, 1974b). The relationship is linear over the temperature range 10-22.5 °C (Figure 11). Outside this range the relationship is not linear and allowance is made for this in the model (Subsection 3.2.4):

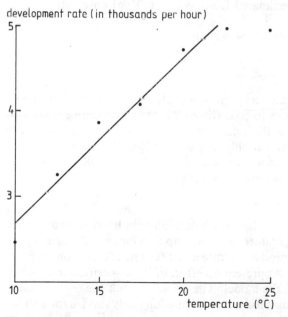

Figure 11. Relationship between developmental rate of cereal aphids and temperature (after Dean, 1974b).

$$DEVELOPMENT\ RATE\ =\ TEMPERATURE\ \times\ 0.000196\ +\ 0.000706$$

for which $r = 0.98$, $p < 0.001$, 4 d.f.

This equation gives a developmental threshold temperature of $-3.6\ °C$. This was used to calculate the mean development times, in $H°$, of the first few instars. It was assumed, based on studies of *Brevicoryne brassicae* (Hughes, 1963), that the fourth instar alatiform aphids take 1.5 times longer to develop than apterous fourth instar aphids, less detailed results for *S. avenae* indicate that this assumption is correct. The length of time that adult aphids survive in the field was based on the assumption that they will not survive as long as those reared in the laboratory (Dean, 1974b). In the model adult longevity is affected by crop developmental stage, although this has not yet been confirmed experimentally. Adult longevity can also be calculated from the cumulative proportion of aphids dying. In most cases longevity is normally distributed and a probability plot linearizes the relationship. The maximum longevity used by Rabbinge (1976) and Rabbinge et. al. (1979) was the average plus three times the standard deviation.

Nymphal survival rates are the mean of the results obtained by Dean (1974b) for a range of temperatures, but reduced after crop developmental stage 73 from 93.8% to 45% (birth to adult moult) for apterous aphids and 93% to 37.4% for alatiform aphids to conform with the observations of Watt (1979a). Adult survival rates used are 90% (adult moult to maximum age) before stage 73 and 60% after this, again estimated from Watt's (1979a) observations.

3.2.3 Reproduction and morph determination

The submodel

It is assumed that alate adults are reproductively mature on arrival in the crop, while apterous adults have to pass through a pre-reproductive delay before reproducing. Figure 12 is the relational diagram for reproduction and morph determination. Thus two variables are used to describe apterous adults: TOTAD – the total number of apterous adults, and TOTADR – the total number of reproductively mature apterous adults.

$$TOTAD \geq TOTADR.$$

Reproductive rate is dependent on the morph of adult aphids, whether apterous or alate (Wratten, 1977), temperature and the crop developmental stage. Apterous aphids have maximum reproductive rate at $20\ °C$ when feeding on crops at developmental stages 59 to 73. At present no effect of intra-specific competition on reproduction is included in the model. The effect of adult age on reproduction is not taken into account as it is assumed that adults only survive for a short period.

The nymphs produced by apterous and alate adults are summed before the proportion that will develop into alates (ALATE) is calculated. Watt & Dixon (1981) have shown experimentally that the offspring of alate adults can develop into alates. The proportion is calculated using a multiple regression equation that incorporates crop developmental stage and aphid density. As aphid density increases and the crop ripens, the proportion of nymphs that develop into alates increases.

The data

Wratten (1977) has shown that there is a marked difference between the reproductive rate of apterous and alate adults. The reproductive rate of apterae, calculated from Dean (1974b), for aphids on barley seedlings (cv. Proctor) is 0.0062 nymphs $H^{\circ -1}$ adult^{-1}; for alates it is 0.0048 nymphs $H^{\circ -1}$ adult^{-1}.

Watt (1979a) has shown that the reproductive rate of S. avenae is 1.6 times higher on the ears of oats than on the leaves. The rates given above were therefore assumed to apply before crop developmental stage 59 and between stages 73 and 83, i.e. milky-ripe stage. Between stages 59 and 73 these rates are multiplied by 1.6, which gives 0.0100 nymphs $H^{\circ -1}$ adult^{-1} for apterae and 0.0079 nymphs $H^{\circ -1}$ adult^{-1} for alates. After crop developmental stage 80 and below 10 °C and above 30 °C the reproductive rate is set to zero. Between 20 and 30 °C the reproductive rate is reduced linearly to zero.

In the model, the morph of the aphid is decided at birth (Dewar, 1977) and is dependent on two variables: aphid density and crop developmental stage. The

Table 3. The proportion of offspring born to six groups of apterous mothers during the first two days and the third to seventh days of reproduction that developed into alatae when reared in groups on the second leaf and ears of wheat (cv. Opal) at various developmental stages.

Group	Proportion alate offspring born on days 1 and 2 and reared on developmental stage 12	Alate offspring born on days 3 to 7	
		developmental stage	proportion
1	0.51	12	0.70
2	0.45	59	0.30
3	0.50	65	0.36
4	0.54	71	0.71
5	0.50	75	0.87
6	0.44	>80	0.94

Ankersmit & Dijkman (to be published).

dependence on crop developmental stage has been shown by Ankersmit & Dijkman (unpublished data; see Table 3) and Watt & Dixon (1981). Multiple regression analysis using the sampling results from two wheat fields near Norwich studied in 1977 was used to determine the relationship between the proportion of alates developing, aphid density and crop developmental stage.

As the duration of the alatiform fourth instar was assumed to be 1.5 times the length of the apterous fourth, the number of the latter was multiplied by 1.5 to compensate for this. The number of alatiform fourths was then expressed as a percentage of the total number of fourth instar aphids present. This was used as the dependent variable.

Aphid density (TOTDEN) in each field was plotted against physiological time on the x axis, with a threshold temperature of -3.6 °C. Crop developmental stage (GSTAGE) was also plotted against accumulated D°, using 6 °C as the threshold temperature. The alatiform fourth instar nymphs recorded in the field were born, on average, 186 D° previously in terms of aphid development and 110 D° in terms of crop development. The difference in time is due to the different threshold temperatures. Thus the aphid density and the crop developmental stage at the birth of these aphids were estimated. These values were then used as the independent variables in the multiple regression:

$$\% \, ALATE = 2.603 \times TOTDEN + 0.847 \times GSTAGE - 27.189$$

for which $r^2 = 0.96$.

Both independent variables are significant (Table 4); this is supported by field observations of Ankersmit (personal communication), Rabbinge et al. (1979) and Watt & Dixon (1981), which show that even for years when aphid density was low a high proportion of alatiform nymphs was produced as the crop ripened.

Table 4. Regression coefficients of the multiple regression relating the proportion of nymphs developing into alatae to the number of aphids per tiller and crop developmental stage.

	Coefficient	S.E.	t-value
Average number of aphids per tiller	2.603	0.340	7.659**
Crop developmental stage	0.847	0.339	2.501*
Constant	-27.189	15.956	-1.704 N.S.

* $p < 0.05$; ** $p < 0.001$; N.S. = not significant.

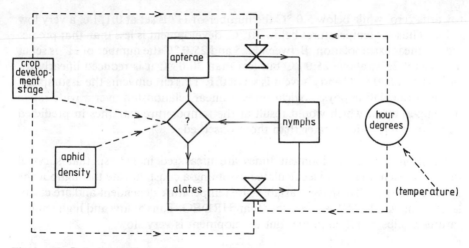

Figure 12. Relational diagram for reproduction and morph determination of cereal aphids.

3.2.4 Physiological processes

The submodel

The most widely used system for a time scale in insect population models is that of physiological time based on accumulated H° or D° above a threshold (Hughes, 1962). This approach is not used here, instead each hour, development, survival and reproduction of the aphids, predator consumption (see Subsection 3.3.3.) and crop development (see Section 3.4) are calculated, based on the prevailing number of H°. In this approach a real-time scale is used, which makes it easier to test the validity of the model's predictions. Thus it is similar to the approach used in other monographs on simulation (e.g. Rabbinge, 1976; Barlow & Dixon, 1980).

Initially, the time of sunrise (to the nearest integer) is calculated as this is the time when temperature is at a minimum. The actual time depends on the latitude of the site and the day number (time of year). Hourly temperatures are calculated by fitting a sine curve through the minimum and maximum (at 1400 h) temperatures for each day. An average over each hour is calculated (AMTEMP) by linear interpolation between the temperatures at the start and end of the hour. For calculating aphid development this average temperature is converted to H° (HRDEG) by the addition of 3.6 over the temperature range 10.0 °C to 22.5 °C. Over this range of temperatures the relationship between development rate and temperature is linear and it is assumed that the aphid's reaction to temperature change is instantaneous (Rabbinge et al., 1979). For average temperatures less than 10 °C but greater than 5.0 °C the number H° is reduced linearly

27

towards zero, while below 5.0 °C the number of H° is set at 0.1, i.e. a very low value. Thus for temperatures below 10 °C, development is less than that predicted by linear extrapolation. Between 22.5 and 25.0 °C the number of H° is set at a constant 26.1; above 25.0 °C but less than 30.0 °C it is reduced linearly towards 0.1; at 30.0 °C and above it is set at 0.1. This circumvents the assumption made in calculating physiological time, a linear relationship over a wide range of temperatures, which would result at the temperature extremes in predicted developmental rates greater than those observed.

Longevities and development times are measured in H°, so that survival rates, for each hour can be calculated assuming a constant rate throughout the aphid's life-span. These two variables are temperature dependent and are calculated using AMTEMP + 3.6 rather than HRDEG. Thus at low and high temperatures aphids continue to die, but development is very slow.

Reproductive rates are measured in nymphs $H°^{-1}$ adult^{-1}, but this process has a different relationship with temperature than development and AMTEMP + 3.6 is used instead of HRDEG. Below 10.0 °C and above 30.0 °C reproduction stops. Between 10.0 and 20.0 °C reproduction increases, but between 20.0 and 30.0 °C it decreases linearly. Reproduction also depends on the adult morph and the crop developmental stage (see Subsection 3.2.3).

The data

The calculation of the time of sunrise follows that used by Rabbinge et al. (1979), but with time as an integer in hours rather than as a real number. The algorithm used to calculate hourly temperatures, using the minimum and maximum values, is an adaptation of that used by Carter (1978) and Rabbinge et al. (1979).

The number of H° is calculated by adding 3.6 to the mean hourly temperature. The adjustments made to the number of H° at the temperature extremes are partly based on Dean's results (1974b) and partly established by trial and error. The adjustments to the number of H° used to calculate reproduction are also based on Dean's results (1974b) and estimates of what is happening at the temperature extremes (i.e. below 10 °C or above 25 °C). It would be useful to know the effect of these temperature extremes, especially whether there is a lag in the resumption of development and reproduction when conditions become more favourable.

3.2.5 Adaptation of the model for other cereal aphid species

The same model structure used for *S. avenae* can be applied to other cereal aphid species, i.e. *M. dirhodum* and *R. padi,* providing the aphid parameters and variables are changed. This has been done for *M. dirhodum* using biological characteristics derived from the laboratory studies of Dean (1973a, 1974b) and Ankersmit (personal communication). The predicted population growth is in good agreement with that observed in different years and different places (Rabbinge & Marinissen, 1981. Internal Report, Department of Theoretical Production Ecology, Agricultural University, Wageningen).

When the model is initiated with suction trap catches the concentration factor determining the number of aphids colonizing a crop needs to be changed. There is little information on the flight behaviour of *M. dirhodum* and *R. padi*. The concentration factors for the two species would have to be determined by comparing predicted and observed numbers of alates. Their threshold temperature for development, and hence development times, are different (e.g. for *M. dirhodum* it is $-2.2.$ °C), as are their reproductive rates and their dependence on temperature and crop developmental stage (Vereijken, 1979; Watt, 1979). Survival rates and their interactions with temperature and crop developmental stage are also likely to vary for each species. Finally, although morph determination is dependent on aphid density (Dixon & Glen, 1971; Elkhider, 1979), the crop developmental stage is also likely to be important and to differ in its effect for each species.

Thus although the structure of the model need not be changed, the biology of each aphid and their response to changes in their environment will have to be determined.

3.3 Natural enemies

3.3.1 Introduction

Three levels of approach can be adopted when simulating the effects of natural enemies: empirical, simplified interaction and complex interaction. The natural enemies included in this study are coccinellids, parasitoids and fungi. Coccinellids are the most common predators in cereal fields in Norfolk (Carter et al., 1980) although in other parts of Europe syrphids may also be numerous, e.g. the Netherlands (Rabbinge et al., 1979). The submodel for coccinellids is based on a simplified interaction, whereas that for the combined effects of parasitoids and fungi is empirical. The numbers of coccinellids, recorded in the field, are used to calculate the number of aphids eaten per hour while the mortality due to parasitoids and fungi, estimated from field counts of mummified and diseased aphids, is used directly in the model. At present the different species of natural enemies are lumped together. Account will be taken of the effects of individual species as more information becomes available. The effects of polyphagous predators are ignored because their searching and consumption rates are unknown. Much of the detailed laboratory information on these predators is not suitable for use in the model.

3.3.2 Parasitoids and diseases

The submodel

The effects of these organisms are included within the main aphid model, operating after the fourth instar aphids have been updated but before the alates emigrate. If PARNO equals one or if there are no parasitoids or disease then this submodel is omitted.

The total number of susceptible adults, TOTFAD, is calculated by summing the numbers in the first age class of the apterous, ADULTS (1), and the alate, ALATED, adults. The number dying per million tillers is PARSIT, the product of the total number of aphids, TOTDEN $\times$ 10^6, and the mortality factor. The numbers of alate (PARAL) and apterous (PARALD) adults killed are calculated, assuming both morphs are equally vulnerable to parasitoids or disease. These are then substracted from their respective age classes. The number of aphids killed by parasitoids and diseases every 24 h is TOTPAR.

The data

Very little research has been done on the population dynamics of the parasitoids and fungi of cereal aphids. As a large number of species is involved this component is very complicated. A study of their interactions with cereal aphids would result in more realistic models.

Counting the number of mummies and diseased aphids in the field underestimates the mortality caused by parasitoids and entomophagous fungi (McLean, 1980), and to correct for this the results should be multiplied arbitrarily by 2 before they are used in the model. Powell (1980) has shown that part of the explanation is that aphids parasitized by *Toxares deltiger* mummify on or in the soil and so are not recorded in the tiller searches.

To get an accurate estimate of the proportion of aphids parasitized or diseased on each sampling occasion it is necessary to have counts of (*a*) live healthy aphids, (*b*) live, but parasitized or diseased aphids and (*c*) dead, mummified or diseased aphids. Then the percentage mortality is:

$$\frac{b + c}{a + b + c} \times 100$$

Unfortunately mortality due to parasitoids and disease is recorded in the field as numbers of mummified and diseased aphids. The proportion of the total population parasitized or diseased is used to calculate the number of newly moulted adults that died each hour. This is an over-simplification but the sub-model cannot be made more realistic until more is known about the biology of the parasitoids and diseases.

3.3.3 Coccinellids

The submodel

The number of aphids in each instar is converted to aphid units (AU) using the following transformation: an aphid unit is equivalent to 1.0 adult, 1.5 fourth instar, 2.0 third instar, 3.5 second instar or 5.0 first instar aphids (Lowe, 1974). The transformation is based on size differences of the various instars. A more accurate method would also take into account the 'attractiveness' of each instar, capture efficiency and prey utilization (Rabbinge, 1976) but this information is not available for coccinellids feeding on *S. avenae*. The aphid units are summed to give the total AFIDUN, and the number of aphid units made up of the first three instars, ONYAFD. The predator subroutine, PREDTR, can be omitted if necessary, i.e. if there are no predators present or if the temperature is less than 15 °C or greater than 30 °C.

If PREDTR is called, the value of ONYAFD is checked. If there are no aphids present in the first three instars (young aphids) then the first part of the sub-routine is skipped. If ONYAFD is not equal to zero then the proportion of the total number of aphid units that are young aphids is calculated, PERPRE. The number of aphid units killed by all predator instars, TOTNY, is calculated from (a) the consumption rates in aphid units per H° for each larval instar of *Coccinella 7-punctata* given in Table 5, and (b) the numbers of larvae in each instar and the number of H° for that instar, assuming the predator has an activity temperature threshold of −3.6 °C. As the fourth instar and adult beetles also

Table 5. Consumption rates of coccinellid larvae at 20 °C when fed excess aphids in the laboratory.

Coccinellid instar	Aphids eaten		
	No (3rd instar) d^{-1}	AU d^{-1}	AU $(H°)^{-1}$
I	4	2.0	0.0053
II	13	6.5	0.0172
III	21	10.5	0.0278
IV	36	18.0	0.0477

After McLean (1980).

eat fourth instar and adult aphids, the number of aphid units made up of the first three instars is reduced by PERPRE. Then the proportions of young aphids killed, PRDFC, and surviving, 'PRDFAC', are calculated.

If there are no young aphids present then fourth instar and adult coccinellids prey only on fourth instar and adult aphids. The number of aphid units killed, TOTADC, is calculated and TOTNY is set to zero and PRDFAC to unity. If there are young aphids present then TOTADC will be lower as the fourth instar and adult beetles will also prey on these. The proportions of fourth instar and adult aphids killed, PRDAD, and surviving, PRDADC, are calculated. Total predation, 'TOTCON', is 'TOTNY' + 'TOTADC'. Control is then returned to the main program.

If ALOWAF (switch variable) equals one and ONYAFD or ADUAFU (= AFIDUN − ONYFAD) are less then 3.0 aphid units, then PRDFAC and PRDADC, respectively, increase linearly as aphid density declines. This means that at low aphid densities coccinellids are not consuming aphids at the maximum rate.

The number in each instar age class is reduced by the appropriate amount, as are the totals for each instar. A new total aphid density is calculated and TOTCON is summed over 24 h, from 1300 to 1200 the following day, to give daily consumption, DAYCON.

The data

The daily consumption of *C. 7-punctata* larvae, fed an excess of third instar aphids at 20 °C in the laboratory, has been determined by McLean (1980). The results are presented in Table 5. The numbers of aphids were converted into aphid units (AU), as outlined above. These experiments were done under conditions of constant temperature and day length, and it was assumed that the coccinellids only fed during the day. The hourly consumption rate is thus total consumption divided by 16 (i.e. hours of daylight), or in AU H$^{°-1}$, hourly consumption divided by 23.6.

Coccinellid searching behaviour is dependent on temperature (McLean, 1980). In the model no aphids are eaten if the average hourly temperature is less than 15 °C or greater than 30 °C. The threshold aphid density is 3 AU per tiller (a 'guestimate'), below which predation is linearly reduced. There is no information on either the effects of temperature extremes or low aphid densities on coccinellid searching behaviour.

The submodel

The major objective of the present study is the explanation of cereal aphid population growth. Therefore, detailed crop growth and development processes are not strictly relevant, but as cereal aphid population growth is dependent on the crop developmental stage (see Chapter 2 and Section 3.2), it is included (Vereijken, 1979; Watt, 1979a).

At the end of each day the crop developmental stage, GSTAGE, is updated. The number of D° is calculated using an algorithm developed by Frazer & Gilbert (1976) with a developmental threshold temperature of 6 °C. The developmental stage is calculated using a polynomial equation, based on the accumulated number of D°, designated TOT.

Rabbinge et al. (1979) used an alternative approach. Crop developmental rate, expressed in decimal units, is linear up to flowering and again after flowering but with a different slope and intercept. Before flowering the relationship is

$$Y = 0.0007\ X - 0.002$$

for which Y = developmental rate and X is temperature.

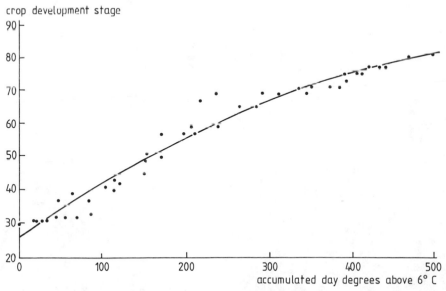

Figure 13. Relationship between wheat development stage and cumulative day degrees above 6 °C in the field (1977).

33

After flowering the relationship is

$$Y = 0.0012\,X - 0.01$$

The latter formula implies that in the period between flowering and end of kernel filling, late-milky-ripe (crop developmental stage 77) 850 °C have been accumulated at 10 °C.

The introduction of a new developmental scale, which is calibrated to give linearity over the whole range of crop developmental stages, could result in the development of a single relationship.

The data

The field results from 1977, from three sites in Norfolk, were used to calculate the polynomial equation used in predicting crop development. Field observations were carried out at least weekly, and usually twice a week after flowering. The resulting equation (see Figure 13) is:

$$\text{GSTAGE} = 0.173 \times TOT - 0.000125\,(TOT)^2 + 26.336$$

for which $r^2 = 0.97$, $n = 48$.

4 Results

4.1 Comparison of model output with observed results

4.1.1 Validation

Verification and validation are terms that are loosely used in simulation studies. In this monograph verification is 'testing to see that the computer program in fact operates on input data in the intended way' (Loomis et al., 1979). Validation is the process of comparing the model's predictions with reality. The cereal aphid model for *S.avenae* has been thoroughly verified, by built-in computer checks and detailed examination of the program.

The simplest means of validation, and probably the most widely used, is a subjective comparison of model predictions with results obtained from field surveys, ideally over a number of years and in different locations. Carter (1978) has criticized many of the early insect population modellers for their failure to validate their models. Teng et al. (1980) use both a subjective comparison and two statistical tests to validate their model, which simulates the development of a barley leaf rust epidemic. The first test is a linear regression analysis of field results against model predictions. However, as the points are in a time series it is not possible to assign significance levels (Barlow & Dixon, 1980). One way to overcome this problem is to compare the observed peak densities, or their timing, with model predictions. These points are independent and the correlation coefficient and the results of the t tests on the slope (should be one) and the intercept (zero) of the relationship give a stringent test of accuracy of the model. The only problem with this approach is a practical one – each point represents a season's sampling of one field. Teng et al. (1980) also use the non-parametric Smirnov test to determine whether there is a significant difference between the observed and predicted values at the point where the two curves differ most. Although this test is less stringent than regression analysis, as it is only testing significance at one point, it does not suffer the drawbacks of time series regression analysis. However, when the incidence of a disease or pest is low the divergence between observed and predicted results might be too small for the test to be meaningful (Teng et al., 1980).

In this study a subjective comparison of population trends and a regression analysis of the observed and predicted population peaks are used to validate the model.

4.1.2 Model predictions

Crop submodel

A comparison of the predictions of the crop development model with field observations (U.K., 1976-1980; the Netherlands, 1976-1979) shows reasonable agreement (Figure 14a-n). The rate at which the crop ripened in the U.K. was overestimated in 1979 (Figure 14f) probably because the wet summer slowed down crop development. At present the effect of rain on crop development is not included in the model. As the U.K. field results for 1977 (Figure 14 c,d) were used to construct the crop development model they are included for reference only. It is not always easy to determine the crop developmental stage, as it can vary within a field and in particular it is difficult to recognize stages within the milky-ripe stage (stage 73-79), which can be prolonged. This is serious because as the grain ripens the survival and reproductive rates and morph determination of the aphid change dramatically and these are primarily responsible for the decline in aphid numbers.

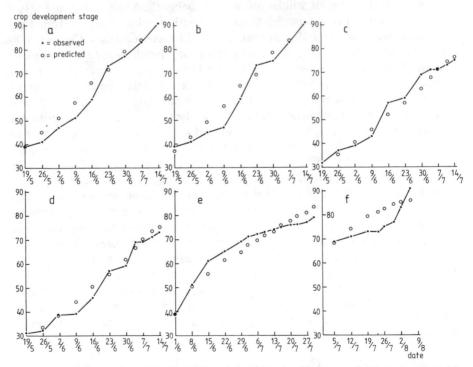

Figure 14. Observed and predicted trends in crop development of wheat for Norwich 1976 (a & b); Norwich 1977 (c & d); Norwich 1978 (e); Norwich 1979 (f); Wageningen 1976 (g); Wageningen 1977 (h); Wageningen 1978 (i); Wageningen 1979 (j); Norwich 1980 (k); Stowmarket 1980 (l); Rothamsted 1980 (m); Brooms Barn 1980 (n).

36

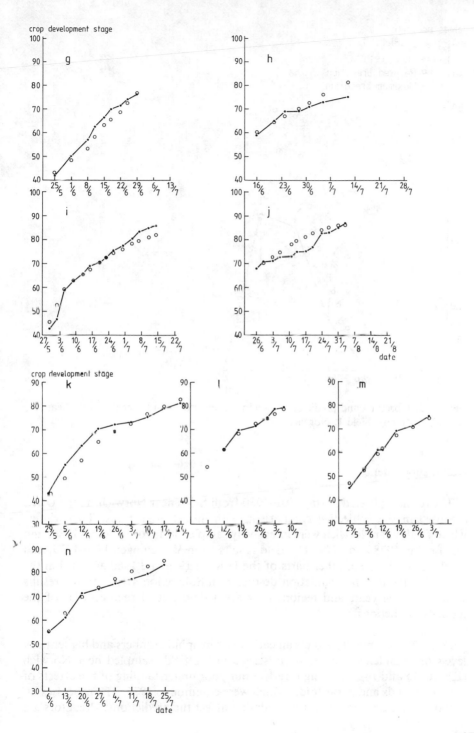

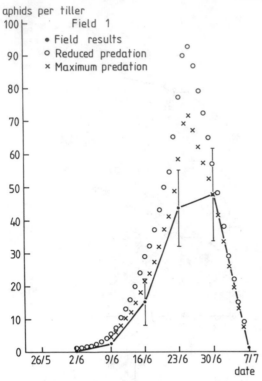

Figure 15. Observed and predicted trends in numbers of aphids per tiller of wheat (cv. Maris Huntsman), Field 1, Norwich 1976.

Aphid submodel

The results collected during 1976-1980 from fields near Norwich, in the U.K., are used in the subjective comparison with model predictions. Two winter-wheat fields near Norwich were intensively sampled in 1976, 1977 and 1980, and one field in 1978 and 1979. The field results from Wageningen 1976-1979, and results collected from other parts of the U.K. in 1979 and 1980 are used in the regression method for validation described in Subsection 4.1.1. As the results varied between years and regions they are a rigid test of the accuracy of the model's predictions.

For 1976, the model predicts an earlier rise in aphid numbers and higher peak levels of abundance than those observed in the fields sampled near Norwich (Figures 15 and 16). This might reflect our poor understanding of the effects of the coccinellids and parasitoids, which were common in 1976. The processes incorporated in the submodel for predation affect the predictions. Predators are

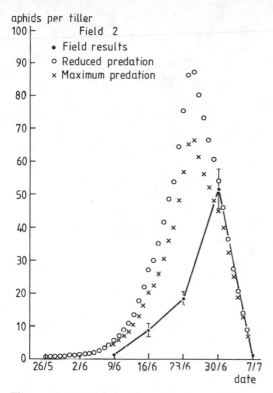

Figure 16. Observed and predicted trends in numbers of aphids per tiller of wheat (cv. Maris Huntsman), Field 2, Norwich 1976.

assumed either to eat the maximum number of aphids possible each day, irrespective of aphid density; or their consumption is reduced linearly at aphid densities less than three aphid units per tiller (Subsection 3.3.3). Maximum consumption by the coccinellids, independent of aphid density, gives a better prediction of the aphid population trends than does reducing predation at low aphid densities. This means that either coccinellids are efficient at finding aphids at low densities or some other factor is missing from the model. As the instar and morph of the aphids were not recorded in 1976 it is difficult to account for the difference between observed and predicted population trends.

The predicted population curves for the two fields sampled in 1977 are similar to those observed (Figures 17 and 18). The difference in aphid numbers between the fields, due to a slight difference in crop developmental stage, is accurately predicted. As a further test of the accuracy of the model's predictions the numbers in specific instars can be compared with those observed. This indicates that the model is giving an acceptable representation of what is happening in the

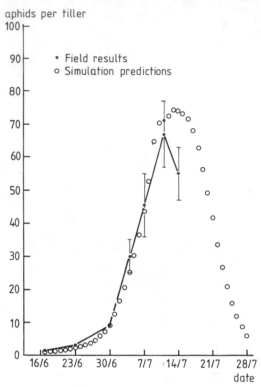

Figure 17. Observed and predicted trends in numbers of aphids per tiller of wheat (cv. Maris Freeman), Norwich 1977.

field (Figures 19 and 20). The number of adult apterae is slightly underestimated (Figure 20) — thus they probably live longer than allowed for in the model. Increase in the time for which adult apterae survive would have to be compensated for by a decrease in the survival rate of the nymphs. Carter (1978) has shown, however, that doubling adult longevity, from 7 to 14 days, at 20 °C, only increases the peak density by 30% and does not affect the date of the peak. Adult longevity needs to be determined under natural conditions. The numbers of immigrant alates observed in the field and those predicted by the model, initialized with data from suction traps, were similar until the end of June. From July onwards the numbers of alates are underestimated (Figure 20). Alatiform fourth instar nymphs are uncommon at this time (when immigration ceases in the model) and hence the increase in the number of alates is due to continued immigration. When a further five days of immigration are included in the model the prediction of alates is closer to the observed but there is little effect on the peak density reached. This is because at this time alates represent less than 10% of the total adult population.

40

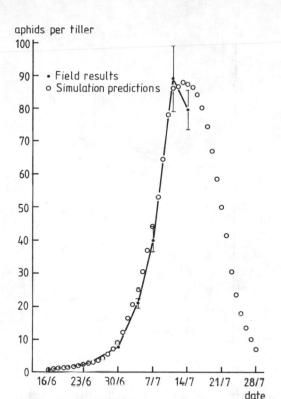

aphids per tiller

Figure 18. Observed and predicted trends in numbers of aphids per tiller of wheat (cv. Maris Huntsman), Norwich 1977.

The predictions for 1978 overestimate the peak density by more than three times: 17.2 per tiller rather than 5.0 (Figure 21). However, the shape of the population curve is accurately predicted. The numbers in all instars are overestimated (Figures 22 and 23). The low temperatures and high rainfall in the summer of 1978 might have reduced the survival rates of nymphs and adults in the field below that included in the model. Natural enemies were uncommon and had little impact on aphid population build-up, although it is possible that the effects of *Entomophthora spp.* were underestimated.

In 1979 *S. avenae* was rare, although *M. dirhodum* was very common. As in 1978 only a few alate *S. avenae* colonized cereals, and they arrived late in the season (Walters, 1982) and hence aphid population development was retarded. The peak density predicted is half that observed (Figure 24), probably because the predicted crop development is too quick. A slower ripening of the crop in the model for 1979 gives aphid populations that are in good agreement with the observed populations (Rabbinge & Marinissen, unpublished results). This illus-

41

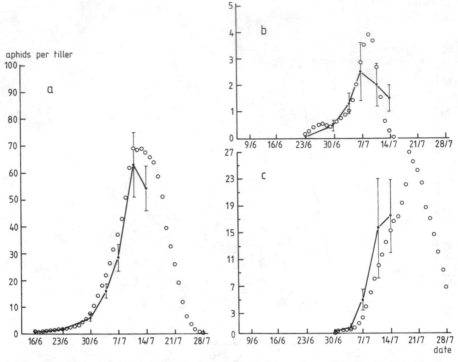

Figure 19. Observed (•) and predicted (○) trends in numbers of (a) first to third instar; (b) apteriform fourth instar; and (c) alatiform fourth instar aphids per tiller of wheat (cv. Maris Huntsman), Norwich 1977.

trates how important the aphid-host interaction is in aphid population dynamics.

For 1980, the model predicts greater numbers of aphids than those observed (Figures 25 and 26). Predation, as in 1976, was important in reducing aphid population levels. The model does predict the difference between the two fields – i.e. aphids achieved a higher density in the field of cv. Maris Huntsman. Weather conditions during June, especially near the end, were adverse as rainfall was high: over 40 mm of rain fell on 30 June. This probably reduced the population growth rate.

Linear regression analysis of the relationship between predicted and observed peak numbers, for the above results and those from other locations, reveals that the model predictions are reasonably accurate (Figure 27). However, the 1980 English field results, other than those for Norwich, are much lower than predicted, as populations declined after flowering. This was due to predators and parasitoids and adverse weather conditions. These results were omitted from the

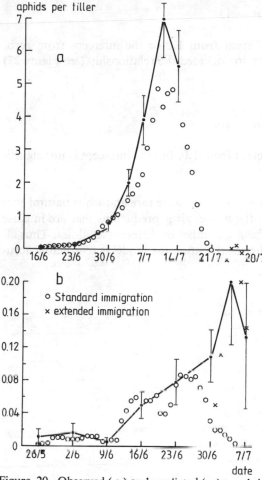

Figure 20. Observed (•) and predicted (○) trends in numbers of (a) apterous adult; and (b) alate adult aphids per tiller of wheat (cv. Maris Huntsman), Norwich 1977.

regression analysis although they are presented in Figure 27. When more is known about the effects of natural enemies it is hoped to improve the predictions.

S. avenae at Rothamsted in 1979 achieved low peak densities − less than 1.0 per tiller. The predicted peaks were also low − less than 2.0 per tiller, although on a log scale this difference is marked. If these 1979 Rothamsted results are omitted from the regression analysis the relationship (see Figure 27) is

$$Y = 0.74 \, X + 0.25$$

43

for which $r = 0.86$, $p < 0.001$, $n = 13$.

The slope is not significantly different from 1.0 or the intercept from zero. When the 1979 Rothamsted results are included the relationship (see Figure 27) is:

$$Y = 1.20 X - 0.53$$

for which $r = 0.87$, $p < 0.001$, $n = 16$.

The slope is not significantly different from 1.0, but the intercept is just significantly less than zero.

With the exception of the years when aphids were rare and when natural enemies were particularly important, the model gives predictions that are in close agreement with observed results from a number of different localities. Thus although some processes need further clarification the model is reasonably accurate.

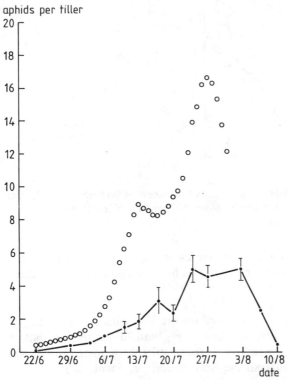

Figure 21. Observed (•) and predicted (○) trends in numbers of aphids per tiller of wheat (cv. Maris Huntsman), Norwich 1978.

44

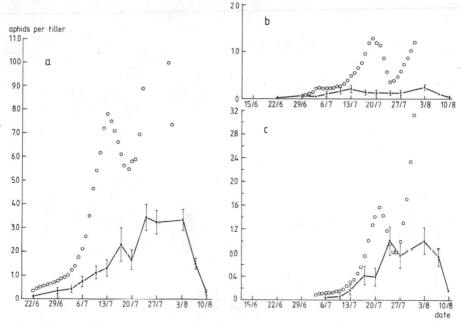

Figure 22. Observed (•) and predicted (○) trends in numbers of (a) first to third instar; (b) fourth instar; and (c) fourth instar alatiform aphids per tiller of wheat (cv. Maris Huntsman), Norwich 1978.

4.2 Sensitivity analysis

4.2.1 Introduction

During and after the construction of a model the importance of its components can be assessed using sensitivity analysis. Carter & Rabbinge (1980) describe two forms of sensitivity analysis: fine, and coarse. Fine is where small positive and negative changes are made to individual components to estimate the level of accuracy needed to obtain reliable results. The size of the changes depends on the accuracy of the components. Those that vary little can only be changed slightly, i.e. within the confidence intervals, whereas those that vary greatly require testing over a considerable range of values, but again within the confidence intervals. When a small change in a component results in a large change in the prediction, a super-proportional response, then the value of that component has to be known accurately. Such changes may also alter the form of the response, i.e. do the changes result in a symmetrical or asymmetrical change in the form of the predictions? In coarse sensitivity analysis the compo-

45

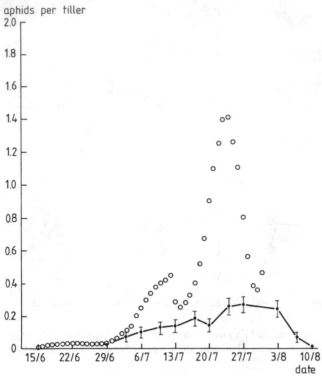

Figure 23. Observed (•) and predicted (○) trends in numbers of aptcrous adult aphids per tiller of wheat (cv. Maris Huntsman), Norwich 1978.

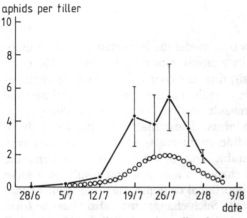

Figure 24. Observed (•) and predicted (○) trends in numbers of aphids per tiller of wheat (cv. Maris Huntsman), Norwich 1979.

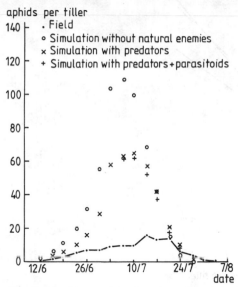

Figure 25. Observed (•) and predicted (○) trends in numbers of aphids per tiller of wheat (cv. Maris Freeman), Norwich 1980.

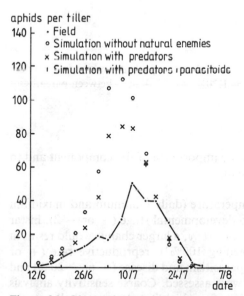

Figure 26. Observed (•)and predicted (○) trends in numbers of aphids per tiller of wheat (cv. Maris Huntsman), Norwich 1980.

47

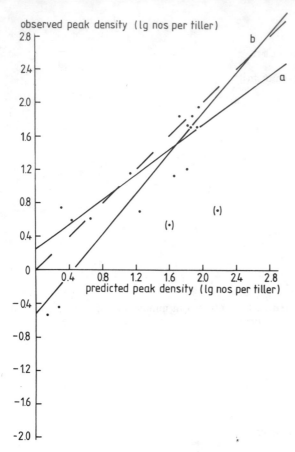

Figure 27. Relationship between predicted and observed peak densities of *S. avenae*. Line a is Y = 0.74X + 0.25; Line b is Y = 1.20X − 0.53. Values between parentheses are omitted from regression analysis.

nent is omitted. This is done to assess the importance of the component and to determine its overall effect in the system.

The effects of small changes to temperature (daily minimum and maximum values + or − 1 °C), the initial crop developmental stage (+ or − 2), instar length (+ or − 20%), survival rate (+ or − 6%, a larger change would result in some instances in survival rates exceeding 100%!), reproductive rate (+ or − 20%), immigration (+ or − 20%), parasitism and disease (+ or − 20%) and morph determination (+ or − 20%) were assessed. Coarse sensitivity analysis was carried out on morph determination, for which it was assumed all nymphs born would develop into apterous adults. Removal of parasitism and disease

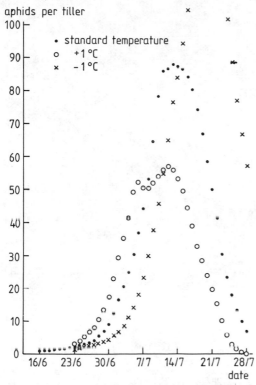

Figure 28. The effect on predicted cereal aphid population trends of small changes in daily minimum and maximum temperatures.

has very little effect in the years studied (Carter, 1978, unpublished). As the predictions for 1977 were close to the observed values, the data of that year were chosen for the sensitivity analysis. As predation was negligible in 1977 it was not included in the sensitivity analysis.

4.2.2. Fine sensitivity analysis

Temperature

A reduction of 1 °C in the daily minimum and maximum temperatures leads to a delay in the date of the peak aphid density of six days, but results in an increase in its size; 153.7 aphids per tiller instead of 87.9 (Figure 28). An increase of 1 °C leads to an earlier — by one day — and lower — 57.0 aphids per tiller —

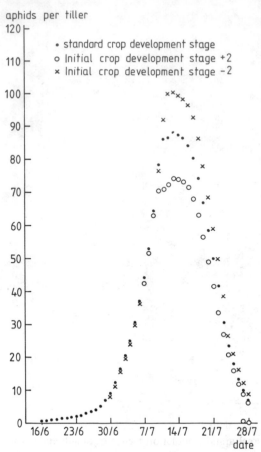

aphids per tiller

- • standard crop development stage
- ○ Initial crop development stage +2
- × Initial crop development stage −2

Figure 29. The effect on predicted cereal aphid population trends of small changes in the initial crop developmental stage.

peak. The reason for this is the differential effect of temperature on crop development and aphid population growth. Low temperatures slow down crop development more than aphid population growth, thus the aphids have more time for reproduction while the crop is optimal. The relationship is complex because not only is the response in peak size asymmetrical but so is its timing. A small change in temperature has a large effect, which indicates that it is an important factor.

Initial crop developmental stage

Changing the initial crop developmental stage affects the size of the peak population density and the response is symmetrical (Figure 29). In previous sen-

50

sitivity analyses changing the initial crop developmental stage resulted in a change in the timing of the peak density as well as its size (Carter, 1978; Carter & Rabbinge, 1980). The major difference between the models they used and this one is that crop development is represented by a polynomial equation rather than linear regression. The linear regression model was less reliable during the critical milky-ripe stage. However, another model using two linear regressions, one before and one after flowering gave reasonable results (Rabbinge ct. al., 1979). The influence of factors other than temperature on crop development need to be studied and a new relationship produced.

Instar length

Alterations to the length of nymphal and adult instars by + or − 20% have a dramatic influence on aphid population growth (Figure 30). An increase in instar length, which extends the pre-reproductive period and adult longevity, delays the peak density by eight days and reduces it from 87.9 to 60 aphids per til-

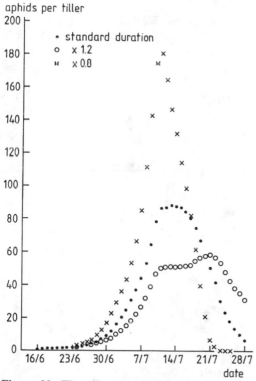

Figure 30. The effect on predicted cereal aphid population trends of small changes in aphid instar duration.

ler (33% decrease). An equivalent decrease in instar length brings the timing of the peak forward two days and increases its size to 180 aphids per tiller (105% increase). This is an example of a super-proportional and asymmetrical response. Therefore, instar length needs to be known accurately as it has an important and complex effect on the system. Rabbinge et al. (1979) have repeated some of Dean's work (1974b) using wheat plants and obtained similar development times. They have also shown that the aphid's response to temperature change is instantaneous. However, the duration of the fourth instar alatiform nymph at different temperatures and the effect of different crop developmental stages on aphid development still need to be determined.

Survival rate

Changes to the survival rates of the nymphal and adult instars have no effect on the general shape or timing of the population curve, but have a large effect on the size of the peak density (Figure 31). The response is super-proportional

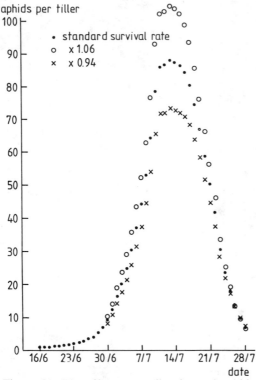

Figure 31. The effect on predicted cereal aphid population trends of small changes in aphid survival rates.

and symmetrical; a 6% change results in a 20% change in the peak aphid population density. Thus survival rates need to be known accurately even though it is extremely difficult to measure them in the field. Few attempts have been made to determine the effects of high winds or torrential rain on aphid survival. The model can be used to obtain estimates, although changing survival rates to improve the prediction, a process called calibration, reduces the general usefulness of the model. Some examples of calibration are described by Loomis et al. (1979). They show that calibration can become a sophisticated curve-fitting routine. Therefore calibration should be omitted or used with care, otherwise an explanatory model may degenerate into a descriptive model.

Reproductive rate

Changes in the reproductive rate, as with survival rate, have no effect on the timing of the peak density, only on its size (Figure 32). The response is super-proportional (Subsection 4.2.1) and asymmetrical as an increase in the reproductive rate leads to a greater absolute change (51% increase) in the peak density than a decrease in the reproductive rate (41% decrease). This means that the reproductive rate needs to be known accurately. Reproductive rate varies with the nutritional status of the host plant, its developmental stage, the age of the aphid and temperature, but these effects have not been determined for wheat. The information available is for other cereals, e.g. effect of host stage of oats

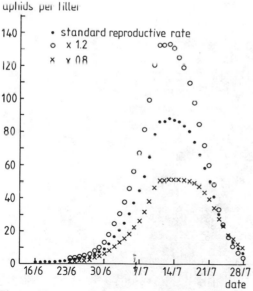

Figure 32. The effect on predicted cereal aphid population trends of small changes in aphid reproductive rates.

on aphid fecundity (Watt, 1979a) and temperature on fecundity of aphids reared on barley (Dean, 1974b). Similar information is needed for different varieties of wheat.

Immigration

As in previous sensitivity analyses (Carter, 1978; Carter & Rabbinge, 1980), when populations are initiated by immigrant alates in early summer, changes in the number of immigrants have a proportional and symmetrical effect on the size of the aphid peak (Figure 33). The timing of the peak is affected only in the simulation where immigration was reduced, but the shapes of the population curves are very similar. Thus the level of immigration has a 1 : 1 effect on the size of the peak density over the range of changes made, and need not be determined as accurately as the survival, developmental and reproductive rates, which have

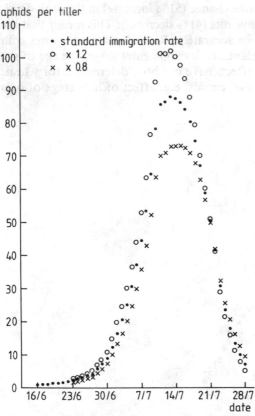

Figure 33. The effect on predicted cereal aphid population trends of small changes in aphid immigration rates.

a super-proportional effect on the peak density. Nevertheless the level of immigration is an important component of the system.

Alate determination

Changing the proportion of nymphs that develop into alate adults affects not only the size of the peak but also its timing (Figure 34). An increase in the proportion leads to a lower peak (79.3 aphids per tiller) two days earlier than the standard simulation (87.4 aphids per tiller); a decrease leads to a higher peak (99.0) two days later. The response is sub-proportional and slightly asymmetrical but this is probably due to the asymmetrical nature of the change. In the standard simulation, if the proportion is greater than 0.83, a 20% increase in the sensitivity analysis simulation results in a proportion greater than 1.0. Thus at proportions greater than 0.83 the resulting percentage change is less than 20%.

The conclusion is that it is not important to know the proportion of alate

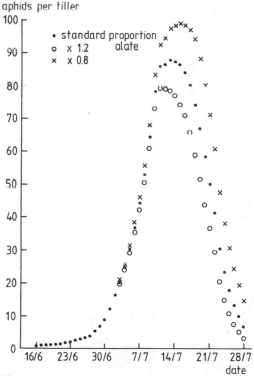

Figure 34. The effect on predicted cereal aphid population trends of small changes in the proportion of alatiform nymphs induced.

nymphs accurately, although the increase in the proportion of alatiform nymphs at the end of the season and the subsequent emigration of the alates is an important factor contributing to the aphid crash.

Parasitism

Similar to previous sensitivity analyses, changing the rate of parasitism has very little effect on either the timing or size of the peak density. It seems likely that the rate of parasitism would have to be high early on in the season to have a significant effect on aphid population increase, and this has not been observed in Norfolk or generally in the Netherlands over the period 1976 to 1980. How-

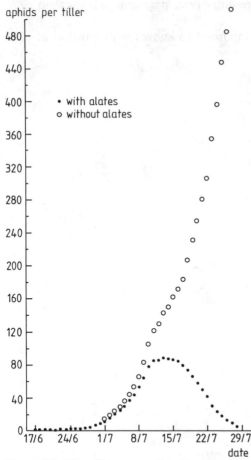

Figure 35. The effect on predicted cereal aphid population trends of omitting alate determination.

ever there are exceptions to this pattern in some places in the Netherlands. There the rate of parasitism was high early in the season and the population growth of the cereal aphids was reduced and resulted in lower aphid peaks (Rabbinge et al., to be published).

4.2.3 Coarse sensitivity analysis

Alate determination

If the production of alate adults, which emigrate from the crop, is omitted this causes a dramatic change in the population development of the aphids (Figure 35). Instead of the aphid population peaking in mid-July, during the milky-ripe stage (stage 73-79), it continues increasing and peaks during the doughy-ripe stage (stage 83-89). As intra-specific competition is not incorporated in the model the effect of removing morph determination from the system is amplified, nevertheless the continued increase in aphid numbers indicates that it is an important component in determining the timing of the aphid crash. This is a general feature of cereal aphid population dynamics in several places in Europe (Watt & Dixon, 1981; Ankersmit et al., to be published) where natural enemies are not very numerous and have not exerted an important influence on the annual aphid population increase over the period 1976-1980.

5 General discussion

5.1 Simulation results

5.1.1 Validation

Although the crop development submodel is purely descriptive it gives a reasonable fit to field observations for a number of different cereal varieties in different localities over a number of years. If new varieties, with markedly different development patterns, e.g. a very short flowering period, were introduced then this would have to be allowed for in the model. The effect of rainfall and soil type on crop development could also be incorporated, but at present this is not thought to be necessary.

The model for *S. avenae* gives predictions that are similar to field observations, except for 1980, when in some localities aphid populations declined after flowering. At present it is not possible to explain why population growth was slow or why the peaks occurred earlier and at lower levels than predicted in 1980. Whether natural enemies can regulate aphid population growth, and under what conditions, is unknown and more work is needed before deciding whether they can be used in the biological control of cereal aphids.

The outbreaks in 1976 and 1977 were accurately described by the model, although the weather conditions in the summers of those two years were very different. The summer of 1976 was much warmer than that of 1977, with a mean June temperature of 16.7 °C compared to 12.3 °C in 1977. May and July temperatures were also higher in 1976 (11.8 °C and 18.0 °C, respectively) than in 1977 (10.1 °C and 15.5 °C). Rainfall was variable within and between the two summers. In June 1976 and July 1977 it was very low (9.3 mm and 3.0 mm, respectively). It could be argued that low rainfall is conducive to aphid outbreaks, because much of the aphid population development occurred in these two months. In 1980, however, high rainfall in June (96.4 mm) and July (51.5 mm) did not prevent *S. avenae* from reaching outbreak levels near Norwich. In 1978 and 1979 the low densities of *S. avenae* on wheat were reasonably accurately predicted. They were the result of a late colonization by a few alate immigrants.

In 1980 the predictions for the Norwich area were reasonably accurate, although the predicted peak densities and population growth were too high. For other areas, further south, and parts of the Netherlands, natural enemies, especially coccinellids, were commoner early on in the season and possibly suppressed the aphid populations, preventing an outbreak. To test this, more information is needed on the interaction between aphids and natural enemies and more specifically with coccinellids and syrphids.

Regression analysis of observed and predicted peak densities is a more stringent test of the model's predictions than qualitative comparisons of observed and predicted aphid densities in successive samples. This latter technique is useful, however, especially when the observed population trend and simulated predictions diverge, because it indicates at what stage the model is wrong. This is especially true if the numbers of aphids present in specific instars are compared.

5.1.2 Sensitivity analysis

Omitting alate determination from the model reveals that when aphid populations reach high levels of abundance it is mainly the increasing proportion of aphids that develop into alate adults and emigrate from the crop that causes the aphid crash at the end of the season. This supports the suggestion that in the absence of natural enemies the induction of alates, together with the lower reproductive and survival rates of the aphids, determine the decline in aphid numbers (Rabbinge et al., 1979; Vereijken, 1979; Watt, 1979; Watt & Dixon, 1981).

The fine sensitivity analysis reveals the level of importance of the different aphid components. Changing the instar length or survival or reproductive rates results in a super-proportional response. Therefore the effect of changes in host plant quality, temperature and other extrinsic variables on these aphid components must be accurately determined. Much of the data used in the model are not for *S. avenae* reared on wheat, but the simulation predictions indicate these values are not unreasonable and are acceptable until more accurate values are available. The values of other components, such as immigration and alate determination, have less effect on aphid abundance. Changing the proportion of aphids developing into alatae results in a sub-proportional response, which is surprising as this component is important in determining the decline in aphid numbers at the end of the season. It is likely, however, that by changing the relationship between alate determination and crop developmental stage a greater response might occur (Rabbinge et al., 1979).

Extrinsic variables, like temperature and initial crop developmental stage, affect the size of the peak aphid density, and temperature the timing of the peak density. A lower temperature somewhat surprisingly leads to a higher peak density but at a later date. This is a consequence of the differential effect of temperature on aphid and plant development (see Rabbinge et al., 1979), which is further complicated by the action of natural enemies. At lower temperatures aphids are present for longer periods of time, so that natural enemies might exert a greater influence on aphid population development (see Vickerman & Wratten, 1979). The earlier (in terms of crop developmental stage) the aphids arrive, the higher the peak density they achieve as they have more time to increase in numbers. The effect of the crop developmental stage on alate settling behaviour is unknown, which is a serious gap in our knowledge of the early phase of aphid population development.

5.2 Simple decision models

5.2.1 Introduction

Carter & Dewar (1981) have reviewed some of the simple decision models that have been developed in an attempt to forecast cereal aphid outbreaks. Suter & Keller (1977) and Vickerman (1977) found that a cold spring was usually followed by a cereal aphid outbreak. Mild springs are thought to allow the build-up of natural enemies early in the year, which then prevent rapid population development of cereal aphids in summer. Cold springs suppress the build-up of natural enemy populations by keeping alternative-prey species scarce without significantly reducing the numbers of cereal aphids. Vickerman (1977) presented several significant relationships between peak aphid populations on cereals in West Sussex in summer and temperature in March and April, e.g. between peak density and lg $(n + 1)$ April air frost days. However, in Norfolk in 1978, the expected cereal aphid outbreak did not occur because very few *S. avenae* colonized cereals whereas in 1980, in Norfolk, an outbreak occurred contrary to the prediction as the natural enemies did not arrive in the cereal crops early enough or in sufficient numbers to reduce significantly aphid population growth. When natural enemies or aphids are scarce the severity of the weather in spring is unlikely to have a marked effect on the subsequent cereal aphid build-up. Therefore the relationship proposed by Vickerman does not apply every year. It might, however, be useful in our present state of understanding as an indicator of potential cereal aphid outbreaks.

Rautapää's (1976) study of forty fields of spring-sown cereals over seven years revealed significant relationships between the peak aphid density of *R. padi* and the numbers present three, seven and 10 days after the first aphids were found. He was also able to show that more coccinellid larvae were present in fields with high densities of *R. padi,* although there was no correlation between either the numbers of first or second generation coccinellid adults and aphid density. However, there was a relationship between the peak aphid population and the number of aphids per coccinellid adult ten days after the first aphids were found. The fewer aphids per coccinellid the lower the resulting peak. Similar relationships were also found for aphids and syrphid larvae although the latter were generally less common than coccinellids. Thus Rautapää was able to predict early in a season the future aphid population trend.

Whalon & Smilowitz (1979) developed a model for the population build-up of the peach potato aphid, *Myzus persicae,* which was used to make short-term predictions of population growth. They assumed exponential population growth dependent on temperature, using data collected in the field from 1975-1977. In 1978, however, their model gave inaccurate predictions. Indeed, early in the season they had to update the simulation predictions to the values observed in the field, otherwise the model would have seriously underestimated population development! Adjusting simulation predictions in this way makes the simulation model redundant.

5.2.2 *Attempts at cereal aphid population prediction in 1980*

Three approaches were used to predict cereal aphid population development: one based on short-term predictions, i.e. three days in advance; and the other two on predicting the peak aphid density. Data were collected from various

Figure 36. Location of the sites sampled in 1980 (•) and the Rothamsted Insect Survey suction traps in England and Wales (×).

places in England (Figure 36): (i) North Farm (near Worthing, Sussex), Hexton (Bedfordshire), Rothamsted (Hertfordshire) and Brooms Barn (Suffolk); (ii) Stowmarket (Suffolk); and (iii) Norwich (Norfolk). The results from (i) and (ii) were used in all three prediction methods, those from (iii) were used only to predict peak populations.

Method 1

The short-term prediction is based on a simplified version of the complex simulation model presented here. Three basic changes were made: the initialization procedure was altered to allow for the introduction of aphid instars in addition to alates; step length was increased to two hours (which only had a minor effect on aphid population build-up); and natural enemies were not considered. Field observations were made once or twice a week. Three-day temperature predictions and general weather conditions were obtained from the Meteorological Office at Bracknell, Berkshire. The model's predictions were sent to the field workers within two days.

Method 2

Significant relationships were obtained when peak densities were plotted against the corresponding aphid densities at different specific crop developmental stages for data collected in Norfolk and at Wageningen, the Netherlands; the later the crop developmental stage the more significant the correlation (Figure 37). This method is similar to that of Rautapää (1976).

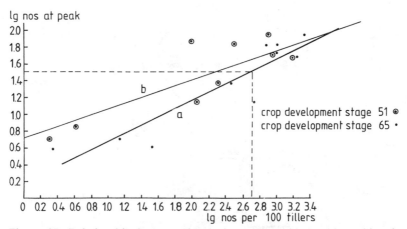

Figure 37. Relationship between the peak cereal aphid densities achieved and cereal aphid numbers at (a) flowering and (b) ear emergence, Norwich and Wageningen 1976-1978.

Crop developmental stage

59    $Y = 0.35 X + 0.71$

for which $r = 0.87$, $n = 10$, $p = 0.001$

65    $Y = 0.50 X + 0.17$

for which $r = 0.92$, $n = 10$, $p < 0.001$
and where Y is lg aphid numbers per tiller at peak and X lg aphid numbers per 100 tillers at the particular crop developmental stage.

Method 3

This method is based on that of Whalon & Smilowitz (1979), but is additionally corrected for crop development. In the simulation model, crop development is determined by accumulated day degrees (above 6 °C) and a similar relationship can be obtained for aphid population development assuming exponential growth (Figure 38):

$$Y = 0.0223 X + 1.7715$$

for which $r = 0.95$, $n = 36$, $p < 0.001$, and where Y is ln aphid numbers per 100 tillers and X is accumulated day degrees.

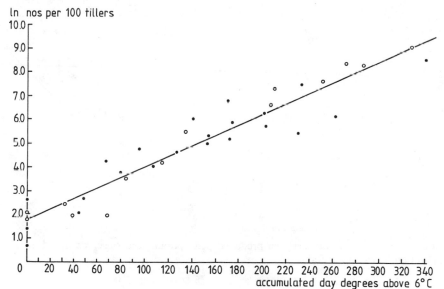

Figure 38. Relationship between the natural logarithm of the numbers of S. *avenae* per 100 tillers and the accumulated day degrees above 6 °C.

The assumption of exponential growth agrees with the observed results. Thus, knowing the aphid density at any particular crop developmental stage and assuming that the peak density occurs at crop developmental stage 75, it is possible to predict the size of the peak population. A three-dimensional diagram has been developed for this purpose (Figure 39). A similar method to this is used in the supervised pest and disease control program (EPIPRE) in the Netherlands (Rabbinge & Carter, in press). In this program, farmers sample fields using simple procedures (Rabbinge & Mantel, 1981) and the results are used to predict the peak density of aphids. This peak density is used to estimate yield loss and help decide whether spraying is justified. This decision is based on a profit-cost analysis that takes into account the costs of spraying, labour needed, etc. (Zadoks et al., in press). This supervised control system is run for about 500 farmers in the Netherlands and may help to lower pesticide usage and to increase profits as the costs of wheat cultivation are lowered (Rijsdijk, Rabbinge & Zadoks, 1981).

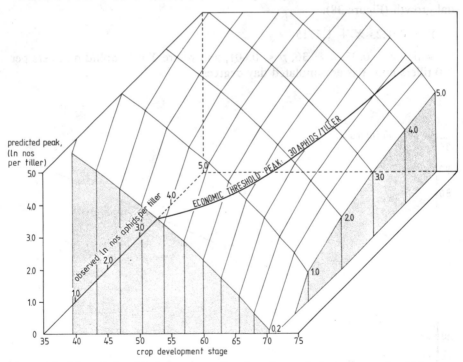

Figure 39. The relationship between predicted peak population, observed aphid numbers and crop developmental stage.

5.2.3 Results and discussion

Unfortunately, aphid population development in eastern England (Figure 36), except near Norwich, was not as predicted by any of the three methods. Population growth was always much slower and the peaks lower and earlier than predicted. This was probably due to a combination of adverse weather conditions and the action of natural enemies, two factors that are not included in the models used for forecasting.

The peak density was accurately predicted for one of the fields sampled in Norfolk, but for another field the observed peak was higher than that predicted by either of the Methods 1 and 2 (Figures 39 and 40). This discrepancy was probably due to the extended ripening period of the crop, which allowed the aphid populations more time to increase.

5.3 Future prospects

These attempts at forecasting cereal aphid outbreaks have revealed many gaps in our knowledge of the cereal aphid system. To define the extent of the system relevant to a warning scheme, a new approach is proposed. The framework of this approach described in Figure 41 is similar to the scheme proposed by Dedryver (unpublished). The life cycles of the aphids, their natural enemies and the crop are studied simultaneously and the interactions between them are emphasised. It is unlikely that this schematic diagram includes all the relevant components in the system. It does, however, include the factors currently thought to be relevant and how those factors interact and affect the likelihood of an outbreak.

The time of sowing affects aphid population build-up, even when aphids do not overwinter on the crop. The relationship is poorly understood but it may be a consequence of earlier sowing, which results in more advanced crop development, and, therefore, less time for the aphids to colonize and increase in numbers on the crops the following spring and summer (sensitivity analysis in Subsection 4.2.2). A reservoir of aphids and natural enemies will develop when aphids overwinter on early-sown crops. The ratio between the two, and the mildness of spring weather, will determine whether the aphids will reach higher levels than on late-sown crops. A study of overwintering aphids, aphid eggs and natural enemies is difficult because they are usually present at very low densities.

In spring, cereal aphids, which colonize cereal fields, may be attacked by large numbers of species of natural enemies. Many of these organisms will also have migrated into cereal crops from habitats where alternative prey may have been numerous. If these natural enemies arrive early enough and in sufficient

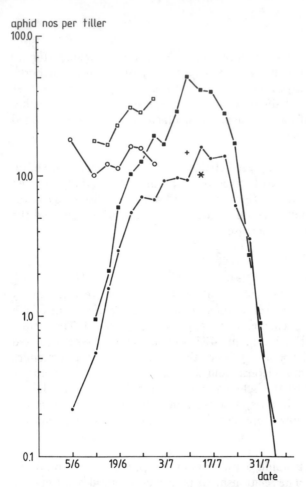

aphid nos per tiller

Figure 40. Weekly predicted peak populations (□, ○) and observed population trends (■, ●) and peak populations predicted from aphid numbers at the end of flowering (∗, +) in two winter wheat fields, respectively, near Norwich, 1980.

numbers, then depending on the number and time of arrival of the immigrant aphids, they might prevent an aphid outbreak. The outcome will also be affected by weather conditions and the level of resistance of the cereal crop to aphids.

For each cereal aphid species and locality it is likely that different parts of the system will be more or less important. Using this framework facilitates the development of effective warning and management schemes on a field to field basis, as for EPIPRE. The simulation models of population growth for the different aphid species, including interactions with their natural enemies and

66

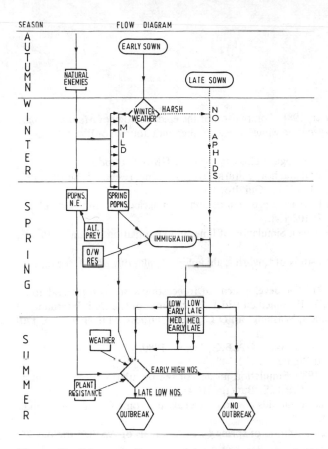

SEASON   FLOW DIAGRAM

Figure 41. Schematic diagram of the seasonal changes in the structure of the cereal aphid ecosystem for use in predicting cereal aphid outbreaks.

weather, will continue to be used to improve understanding of the system. However, because of the complexity of these models it is unlikely that they will be used for prediction. It is more likely that the simpler and more easily applied summary models and decision rules based on the insight gained from the complex models will be used for management and warning programmes.

Acknowledgements

We are grateful to Gert Ankersmit, Herman Dijkman, Geoff. Heathcote, Alan Dewar, Bob Kowalski and the U.E.A. Cereal Aphid Group for permission to use their field results, and the Rothamsted Insect Survey for the suction trap data. Nick Carter is indebted to NATO/S.E.R.C./The Royal Society and the Agricultural Research Council for supporting his part in this work.

References

Ankersmit, G.W. & N. Carter, 1981. Comparison of the epidemiology of *Metopolophium dirhodum* and *Sitobion avenae* on winter wheat. Netherlands Journal of Plant Pathology 87: 71-81.

Anonymous, 1943. The meteorological glossary, 3rd ed. HMSO, London.

Anonymous, 1980. Agricultural and horticultural returns − final results of the June 1979 census, crops and grass, M.A.F.F., Guildford.

Baranyovits, F., 1973. The increasing problem of aphids in agriculture and horticulture. Outlook on Agriculture 7: 102-108.

Barlow, N. & A.F.G. Dixon, 1980. Simulation of lime aphid population dynamics. Pudoc, Wageningen, 171 pp.

Carter, N., 1978. Simulation study of English grain aphid populations. Ph.D. thesis, University of East Anglia.

Carter, N. & A. Dewar, 1981. The development of forecasting systems for cereal aphid outbreaks in Europe. In: T. Kommedahl (Ed.): Proceedings of the IXth International Congress of Plant Protection, Washington, D.C., 1979. Burgess, Minneapolis. p. 170-173.

Carter, N., I.F.G. McLean, A.D. Watt & A.F.G. Dixon, 1980. Cereal aphids − a case study and review. Applied Biology 5: 271-348.

Carter, N. & R. Rabbinge, 1980. Simulation models of the population development of *Sitobion avenae*. I.O.B.C./W.P.R.S. Bulletin III/4: 93-98.

Dean, G.J., 1973a. Bionomics of aphids reared on cereals and some Gramineae. Annals of Applied Biology 73: 127-135.

Dean, G.J., 1973b. Aphid colonization of spring cereals. Annals of Applied Biology 75: 183-193.

Dean, G.J., 1974a. The overwintering and abundance of cereal aphids. Annals of Applied Biology 76: 1-7.

Dean, G.J., 1974b. The effect of temperature on the cereal aphids *Metopolophium dirhodum* (Wlk.), *Rhopalosiphum padi* (L.) and *Macrosiphum avenae* (F.) (Hem., Aphididae). Bulletin of Entomological Research 63: 401-409.

Dean, G.J., 1978. Observations on the morphs of *Macrosiphum avenae* and *Metopolophium dirhodum* on cereals during the summer and autumn. Annals of Applied Biology 89: 1-7.

Dean, G.J., A.M. Dewar, W. Powell & N. Wilding, 1980. Integrated control of cereal aphids. I.O.B.C. Bulletin III/4: 30-47.

Dean, G.J. & N. Wilding, 1971. *Entomophthora* infecting the cereal aphids *Metopolophium dirhodum* and *Sitobion avenae*. Journal of Invertebrate Pathology 18: 169-176.

Dean, G.J. & N. Wilding, 1973. Infection of cereal aphids by the fungus *Entomophthora*. Annals of Applied Biology 74: 133-138.

Dedryver, C.A., 1978. Biologie de pucerons des cereales dans l'Ouest de la France. Annales de Zoologie − Ecologie d'Animaux 10: 483-505.

Dewar, A.M., 1977. Morph determination and host alternation in the apple-grass aphid *Rhopalosiphum insertum* Wlk. Ph.D. thesis, University of Glasgow.

Dixon, A.F.G., 1959. An experimental study of the searching behaviour of the predatory beetle *Adalia decempunctata* (L.). Journal of Animal Ecology 28: 259-281.

Dixon, A.F.G., 1971. The life-cycle and host preferences of the bird cherry-oat aphid, *Rhopalosiphum padi* L. and their bearing on the theories of host alternation in aphids. Annals of Applied Biology 68: 135-147.

Dixon, A.F.G., 1977. Aphid ecology: life cycles, polymorphism and population regulation. Annual Review of Ecology and Systematics 8: 329-353.

Dixon, A.F.G. & A.M. Dewar, 1974. The time of determination of gynoparae and males in the bird cherry-oat aphid *Rhopalosiphum padi*. Annals of Applied Biology 78: 1-6.

Dixon, A.F.G. & D.M. Glen, 1971. Morph determination in the bird cherry-oat aphid *Rhopalosiphum padi* L. Annals of Applied Biology 68: 11-21.

Dunn, J.A. & D.W. Wright, 1955. Overwintering egg populations of the pea-aphid in East Anglia. Bulletin of Entomological Research 46: 389-392.

Elkhider, E.M., 1979. Studies on the environmental control of polymorphism in the rose-grain aphid, *Metopolophium dirhodum*. Ph.D. thesis, Queen Elizabeth College, London.

Fletcher, K.E. & R. Bardner, 1969. Cereal aphids on wheat. Report for Rothamsted Experimental Station 1968, p. 200-201.

Fransz, H.G., 1974. The functional response to prey density in an acarine system. Pudoc, Wageningen, 149 pp.

Frazer, B.D. & N. Gilbert, 1976. Coccinellids and aphids: a quantitative study of the impact of adult ladybirds preying on field populations of pea aphids. Journal of Entomological Society British Columbia 73: 33-56.

George, K.S., 1974. Damage assessment aspects of cereal aphid attack in autumn- and spring-sown cereals. Annals of Applied Biology 77: 67-74.

George, K.S., 1975. The establishment of economic damage thresholds with particular references to cereal aphids. Proceedings of the 8th British Insecticide and Fungicide Conference. the Boots Company Ltd, Nottingham, p. 79-85.

George, K.S. & R. Gair, 1979. Crop loss assessment on winter wheat attacked by the grain aphid, *Sitobion avenae* (F.). Plant Pathology 28: 143-149.

Goudriaan, J., 1973. Dispersion in simulation models of population growth and salt movement in the soil. Netherlands Journal of agricultural Science 21: 269-281.

Heathcote, G.D., 1970. The abundance of grass aphids in Eastern England as shown by sticky trap catches. Plant Pathology 19: 87-90.

Heathcote, G.D., 1978. Coccinellid beetles on sugar beet in Eastern England. Plant Pathology 27: 103-109.

Helle, W. & M. van de Vrie, 1974. Problems with spider mites, Outlook on Agriculture, 8: 119-125.

Hille Ris Lambers, D., 1947. Contributions to a monograph of the Aphididae of Europe. Temminckia 7: 179-379.

Hughes, R.D., 1962. A method for estimating the effects of mortality on aphid populations. Journal of Animal Ecology 31: 389-396.

Hughes, R.D., 1963. Population dynamics of the cabbage aphid *Brevicoryne brassicae* (L.). Journal of Animal Ecology 32: 393-424.

Jeffers, J.N.R., 1978. An introduction to systems analysis: with ecological applications. Edward Arnold, London, 198 pp.

Jones, M.G., 1972. Cereal aphids, their parasites and predators caught in cages over oat and winter wheat crops. Annals of Applied Biology 72: 13-25.

Kolbe, W., 1969. Studies on the occurrence of different aphid species as the cause of cereal yield and quality losses. Bayer Pflanzenschutz Nachrichten 22: 171-204.

Kolbe, W., 1970. Further studies on the reduction in yield by aphid infestation. Bayer Pflanzenschutz Nachrichten 23: 144-162.

Large, E.C., 1954. Growth stage of cereals, illustration of the Feekes scale. Plant Pathology 3: 128-129.

Lamb, H.H., 1973. Climate change and foresight in agriculture: the possibilities of long term weather advice. Outlook on Agriculture 7: 203-210.

Latteur, G., 1976. Les pucerons des céreales: biologie nuisance, ennemis. Memoire no. 3, Centre de Recherches Agronomiques de l'Etat, Gembloux.

Leather, S.R., 1980. Egg survival in the bird cherry-oat aphid *Rhopalosiphum padi*. Entomologia experimentalis et applicata 27: 96-97.

Leather, S.R. & A.F.G. Dixon, 1981. The effect of cereal growth stage and feeding site on the reproductive activity of the bird cherry aphid, *Rhopalosiphum padi*. Annals of Applied Biology 97: 135-141.

Loomis, R.S., R. Rabbinge & E. Ng, 1979. Explanatory models in crop physiology. Annual Review of Plant Physiology 30: 339-367.

Lowe, H.J.B., 1974. Effects of *Metopolophium dirhodum* on spring wheat in the glasshouse. Plant Pathology 23: 136-140.

Maas, S.J. & G.F. Arkin, 1980. TAMW: a wheat growth and development simulation model. A publication of The Texas Agricultural Experiment Station, College Station, Texas, United States, Program and Documentation No. 80-3.

McLean, I.F.G., 1980. Ecology of the natural enemies of cereal aphids. Ph.D. thesis, University of East Anglia.

Markkula, M., 1978. Pests of cultivated plants in Finland. Annales Agriculturae Fenniae 17: 32-35.

Potts, G.R., 1977. Some effects of increasing the monoculture of cereals. In: J.M. Cherrett & G.R. Sager (Eds.): 18th Symposium, British Ecological Society, Origins of pest, parasite, disease and weed problems, Bangor, 1976. Blackwell, Oxford, p. 183-202.

Potts, G.R. & G.P. Vickerman, 1974. Studies on the cereal ecosystem. Advances in Ecological Research 8: 107-197.

Powell, W., 1980. *Toxares deltiger* (Haliday) (Hymenoptera: Aphidiidae) parasitizing the cereal aphid, *Metopolophium dirhodum* (Walker) (Hemiptera: Aphididae), in Southern England: a new host: parasitoid record. Bulletin of Entomological Research 70: 407-409.

Rabbinge, R., 1976. Biological control of fruit-tree red spider mite. Pudoc, Wageningen, 234 pp.

Rabbinge, R., 1981. Warning systems, Proceedings of Symposia IX International Congress of Plant Protection, pp. 163-167.

Rabbinge, R. & W.P. Mantel, 1981. Monitoring for cereal aphids in winter wheat. Netherlands Journal of Plant Pathology 87: 25-29.

Rabbinge, R., G.W. Ankersmit & G.A. Pak, 1979. Epidemiology and simulation of population development of *Sitobion avenae* in winter wheat. Netherlands Journal of Plant Pathology 85: 197-220.

Rabbinge, R. & F.H. Rijsdijk, 1982. EPIPRE: A disease and pest management system for winter wheat taking account of micrometeorological factors. EPPO bull. (in press).

Rabbinge, R. & M.W. Sabelis, 1980. Systems analysis and simulation as an aid to the understanding of acarine predator-prey systems. In: A.K. Minks & P. Gruys (Eds): Integrated Control of insect pests in the Netherlands. Pudoc, Wageningen. p. 281-290.

70

Rabbinge, R., E.M. Drees, M. van der Graaf, F.C.M. Verberne & A. Wesselo, 1981. Damage effects of cereal aphids in wheat. Netherlands Journal of Plant Pathology 87: 217-232.

Rautapää, J., 1976. Population dynamics of cereal aphids and methods of predicting population trends. Annales Agriculturae Fenniae 15: 272-293.

Rijsdijk, F.H., R. Rabbinge & J.C. Zadoks, 1981. A system approach to supervise control of pests and diseases of wheat in the Netherlands. In: T. Kommedahl (Ed.) Proceedings of the IXth International Congress of Plant Protection, Washington, D.C., 1979. Burgess, Minneapolis.

Schneider, F., 1969. Bionomics and physiology of aphidophagus Syrphidae. Annual Review of Entomology 19: 103-124.

Seligman, N.G. & H. van Keulen, 1981. PAPRAN: a simulation model of annual pasture production limited by rainfall and nitrogen. In: M.J. Frissel & J.A. van Veen (Eds): Simulation of nitrogen behaviour of soil-plant systems. Pudoc, Wageningen. p. 192-221.

Sparrow, L.A.D., 1974. Observations on aphid populations on spring-sown cereals and their epidemiology in South-East Scotland. Annals of Applied Biology 77: 79-84.

Sunderland, K.D., 1975. The diet of some predatory arthropods in cereal crops. Journal of Applied Ecology 12: 507-515.

Sunderland, K.D., D.L. Stacey & C.A. Edwards, 1980. The role of polyphagous predators in limiting the increase of cereal aphids in winter wheat. I.O.B.C./W.P.R.S. Bulletin III/4: 85-91.

Suter, V.H. & S. Keller, 1977. Ökologische Untersuchungen an Feldboulich wichtiger Blattlausarten als Grundlage für eine Befallsprognose. Zeitschrift für angewandte Entomologie 83: 371-393.

Taylor, L.R. & J.M.P. Palmer, 1972. Aerial sampling. In: H.F. van Emden (Ed.): Aphid technology. Academic Press, London. p. 189-234.

Teng, P.S., M.J. Blackie & R.C. Close, 1980. Simulation of the barley leaf rust epidemic: structure and validation of BARSIM-I. Agricultural Systems 5: 85-103.

Vereijken, P.H., 1979. Feeding and multiplication of three cereal aphid species and their effect on yield of winter wheat. Agricultural Research Report 888. Pudoc, Wageningen, 58 pp.

Vickerman, G.P., 1977. Monitoring and forecasting insect pests of cereals. Proceedings 1977 British Crop Protection Conference on Pests and Diseases. British Crop Protection Council Publication, Croydon, p. 227-234.

Vickerman, G.P. & K.D. Sunderland, 1975. Arthropods in cereal crops: nocturnal activity, vertical distribution and aphid predation. Journal of Applied Ecology 12: 755-766.

Vickerman, G.P. & S.D. Wratten, 1979. The biology and pest status of cereal aphids (Hemiptera: Aphididae) in Europe: a review. Bulletin of Entomological Research 69: 1-32.

Walters, K.F.A., 1982. Flight behaviour of cereal aphids. Ph.D. thesis, University of East Anglia.

Watt, A.D., 1979a. The effect of cereal growth stages on the reproductive activity of *Sitobion avenae* and *Metopolophium dirhodum*. Annals of Applied Biology 91: 147-157.

Watt, A.D., 1979b. Life history strategies of cereal aphids. Ph.D. thesis, University of East Anglia.

Watt, A.D. & A.F.G. Dixon, 1981. The effect of cereal growth stages and crowding of aphids on the induction of alatae in *Sitobion avenae* and its consequences for population growth. Ecological Entomology 6: 441-447.

Way, M.J. & C.J. Banks, 1964. Natural mortality of eggs of the black bean aphid *Aphis fabae* on the spindle tree *Euonymus europaeus*. Annals of Applied Biology 54: 255-267.

Way, M.J. & M.E. Cammell, 1974. The problem of pest and disease forecasting possibilities and limitations as exemplified by work on the bean aphid *Aphis fabae*. Proceedings 7th British Insecticide and Fungicide Conference. The Boots Company Ltd, Nottingham, 3: 933-954.

Welch, S.M. & B.A. Croft, 1979. The design of biological monitoring systems for pest management. Pudoc, Wageningen, 84 pp.

Whalon, M.E. & Z. Smilowitz, 1979. GPA-CAST, a computer forecasting system for predicting populations and implementing control of the green peach aphid on potatoes. Environmental Entomology 8: 908-913.

Wit, C.T. de & J. Goudriaan, 1978. Simulation of ecological processes. Pudoc, Wageningen, 103 pp.

Wratten, S.D., 1975. The nature and the effects of the aphids *Sitobion avenae* and *Metopolophium dirhodum* on the growth of wheat. Annals of Applied Biology 79: 27-34.

Wratten, S.D., 1977. Reproductive strategy of winged and wingless morphs of the aphids *Sitobion avenae* and *Metopolophium dirhodum*. Annals of Applied Biology 85: 319-331.

Zadoks, J.C., T.T. Chang & C.F. Konzak, 1974. A decimal code for the growth stages of cereals. Eucarpia Bulletin 7, 11 pp.

Zuniga, F. & H. Suzuki, 1976. Ecological and economic problems created by aphids in Latin America. Outlook on Agriculture 8: 311-319.

Appendix A

Program listing of the population model of the English grain aphid
(Sitobion avenae)

```
C A POPULATION MODEL OF SITOBION AVENAE ON WINTERWHEAT
C
C THIS MODEL SIMULATES THE POPULATION GROWTH OF S AVENAE TAKING INTO
C ACCOUNT THE EFFECT OF PREDATORS, PARASITES AND THE HOST PLANT.
C
C
C********1......INITIALISATION*************
C
C
C VARIABLE NAMES WILL BE EXPLAINED AS THEY APPEAR IN THE PROGRAM
C
      REAL NEWNY,NWNY,LAT,MXTT,MNTT
      DIMENSION PRED(250,5),PARA(250),HRDALD(750),ALATAD(750),
     1HRDAD(750),ADULTS(750),FNYMPH(200),HRDFN(200),ALFN(200),
     2HRDALF(200),TNYMPH(150),HRDTN(150),ALTN(150),HRDALT(150),
     3PNYMPH(150),HRDPN(150),ALPN(150),HRDALP(150),SNYMPH(150),
     4HRDSN(150),ALSN(150),HRDALS(150),IMM(200),MNTT(250),
     5MXTT(250),DAYL(250),IRISE(250),RISE(250),TEMP(24),
     6AMTEMP(24),HRDEG(24)
C
C ZERO ALL INSTARS NOS AND AGES STATING WITH THE ADULTS
C
      DO 1 I=1,750
      HRDALD(I)=0.0
      ALATAD(I)=0.0
      HRDAD(I)=0.0
      ADULTS(I)=0.0
    1 CONTINUE
C
C NOW THE FOURTH INSTAR, APTERAE THEN ALATIFORM
C
      DO 2 I=1,200
      FNYMPH(I)=0.0
      HRDFN(I)=0.0
      ALFN(I)=0.0
      HRDALF(I)=0.0
    2 CONTINUE
C
C NOW THE OTHER INSTARS
C
      DO 3 I=1,150
      TNYMPH(I)=0.0
      HRDTN(I)=0.0
      ALTN(I)=0.0
      HRDALT(I)=0.0
      SNYMPH(I)=0.0
      HRDSN(I)=0.0
      ALSN(I)=0.0
      HRDALS(I)=0.0
      PNYMPH(I)=0.0
      HRDPN(I)=0.0
      ALPN(I)=0.0
```

```
              HRDALP(I)=0.0
  3     CONTINUE
C
C ZERO ALL INSTAR TOTALS
C
        TOTADR=0.0
        TOTAD=0.0
        TOTALA=0.0
        TOTFOR=0.0
        TOTALF=0.0
        TOTTHI=0.0
        TOTALT=0.0
        TOTSEC=0.0
        TOTALS=0.0
        TOTFIR=0.0
        TOTALP=0.0
C
C ZERO PREDATOR AND PARASITE ARRAYS
C
        DO 4 I=1,250
        DO 5 J=1,5
        PRED(I,J)=0.0
  5     CONTINUE
  4     CONTINUE
        DO 6 I=1,250
        PARA(I)=0.0
  6     CONTINUE
C
        TOTCON=0.0
        PRDFAC=1.0
        PRDADC=1.0
        DAYCON=0.0
        TOTPAR=0.0
C
C
C**********2....DATA INPUT******************
C
C
C THE FIRST TWO, THREE DIGIT INTEGER NOS ARE THE START AND FINISH
C DAY, DAY 1 IS JAN 1ST.
C
C

        READ(1,7)ISTART,IFINIS
  7     FORMAT(2I3)
C
C NOW THE SENSITIVITY ANALYSIS FACTORS, SEN1=TEMP+/-1 OR 0.
CSEN2=GSTAGE+/-2, SEN3=INSTAR LENGTHX 1.0.8,1.2, SEN4=SURVIVAL
C X .94,1.06,1, SEN5=REPRODUCTION * 1.2,0.8,1, SEN6=IMMIGRATION
C * 1.2,0.8,1, SEN7= PREDATION * 1.2,0.8,1, SEN8= PARASITISM
C *1.2,0.8,1, SEN9= ALATE NYMPHS* 1.2,0.8,1.
C
        READ(1,13)SEN1,SEN2,SEN3,SEN4,SEN5,SEN6,SEN7,SEN8,SEN9
 13     FORMAT(9F5.2)
C
C
C NOW THE TEMPERATURE DATA, ALL THE MAXIMUM DAILY TEMPS,STARTING
C AT DAY ISTART-1 AND FINISHING IFINIS. THEN THE MINIMUM TEMPS
C STARTING AT DAY ISTART AND FINISHING DAY IFINIS+1.
C ONE REAL NUMBER PER LINE
C
```

```
      DO 8 I=ISTART-1,IFINIS
      READ(1,9)MXTT(I)
  9   FORMAT(F10.2)
C
C SENSITIVITY ANALYSIS
C
      MXTT(I)=MXTT(I)+(SEN1)
  8   CONTINUE
      DO 10 I=ISTART,IFINIS+1
      READ(1,11)MNTT(I)
 11   FORMAT(F10.2)
C
C SENSITIVITY ANALYSIS
C
      MNTT(I)=MNTT(I)+(SEN1)
 10   CONTINUE
C
C NOW 3 VARIABLES, LAT= LATITUDE OF THE SITE  GSTAGE=
C THE INITIAL DEVELOPMENT STAGE OF THE CROP,TOT=ACC
C DAY DEG AND TILERS=TILLERS/SQ.M.
C
      READ(1,12)LAT,GSTAGE,TOT,TILERS
 12   FORMAT(4F10.2)
      GSTAGE=GSTAGE+(SEN2)
      TAYPAL=9600.0/(0.4047*TILERS)
C
C NOW THE IMMIGRATION DATA- DAILY SUCTION TRAP CATCHES,
C FIRST THE START AND FINISH DAYS OF MIGRATION AND THE
C CONCENTRATION FACTOR, NORMALLY 40. ALL THREE
C DIGIT INTEGERS.
C
      READ(1,14)IMSTAR,IMFINI,INCONF
 14   FORMAT(3I3)
C
C NOW THE RAW DATA, TEN 3-DIGIT INTEGERS PER LINE
C
      READ(1,15)(IMM(I),I=IMSTAR,IMFINI)
 15   FORMAT(10I3)
C
C NOW THE NATURAL ENEMIES
C FIRST THE TWO NUMBERS, IF EQUAL TO 1, WILL SKIP ROUND THE
C PREDATOR AND PARASITE PROCEDURES. THE THIRD NUMBER IF EQUAL
C TO 1 REDUCES PREDATION AT LOW APHID DENSITIES
C
      READ(1,16)PREDNO,PARNO,ALOWAF
 16   FORMAT(3F10.2)
      IF(PREDNO.EQ.1.0)GO TO 19
C
C
C NOW THE PREDATORS, FIRST THE START AND FINISH DAYS
C
      READ(1,17)IPRED,IPRFIN
 17   FORMAT(2I3)
C
C NOW THE MATRIX, INSTARS 1-5 SEPERATE (5 NOS/LINE) APPROX
C NOS/TILLER
C
      DO 18 I=IPRED,IPRFIN
      READ(1,20)(PRED(I,J),J=1,5)
 20   FORMAT(5F10.4)
 18   CONTINUE
```

75

```
C
C SKIP LINE
C
   19    CONTINUE
C
         IF(PARNO.EQ.1.0)GO TO 21
C
C NOW THE PARASITES FIRST, START AND FINISH DAYS
C
         READ(1,22)IPARA,IPAFIN
   22    FORMAT(2I3)
C
C NOW THE MATRIX, 5 NOS PER LINE, EACH THE HOURLY MORTALITY FOR ONE
C DAY
C
         READ(1,23)(PARA(I),I=IPARA,IPAFIN)
   23    FORMAT(5F10.4)
   21    CONTINUE
C
C THIS FINISHES DATA INPUT
C
C
C AS THE MODEL IS UPDATED HOURLY TEMPS HAVE TO BE CALCULATED
C HOURLY, BUT FIRST THE TIME OF SUNRISE IS CALCULATED(IRISE)
C
         PI=3.1415927
         RAD=PI/180.0
         CONV=RAD*LAT
C
         DO 1234 IDAYY=ISTART,IFINIS+1
         DEC=-23.4*COS(PI*(IDAYY+10.173)/182.621)
         SSIN=SIN(CONV)*SIN(RAD*DEC)
         CCOS=COS(CONV)*COS(RAD*DEC)
         TT=SSIN/CCOS
         AS=ASIN(TT)
         DAYL(IDAYY)=12.0*(PI+2.0*AS)/PI
         RISE(IDAYY)=12.0-(DAYL(IDAYY)/2.0)+0.5
         IRISE(IDAYY)=RISE(IDAYY)
 1234 CONTINUE
C
C THE HEADINGS ARE NOW PRINTED OUT
C
         WRITE(2,600)
   600   FORMAT(1H1,109HIDAYY  I-APT     II-APT  III-APT    IV-APT    V-APT
        1        I-ALT    II-ALT   III-ALT    IV-ALT      V-AL TOTYN//)
         WRITE(2,601)
   601   FORMAT(107HGSTAGE  REP-AD    ALTIM    TOTALE   TOTAL DENSITY   T
        2 0TPAR   DAILY CON   PRDFAC  PRDADC   AFIDUN  TOTDDG///)
C
C
C
C********MODEL NOW STARTS**********
C
C
         DO 107 IDAYY=ISTART,IFINIS
C
C*******3......HOURLY TEMPERATURES ARE CALCULATED*************
C
C
         DO 4320 IT=1,24
         IF(IDAYY.EQ.ISTART.AND.IT.EQ.1)TEMP(24)=((MXTT(IDAYY-1)-MNTT(IDAYY
```

76

```
     1))*(COS(PI*(-10)/(10+IRISE(IDAYY)))))/2.0+(MXTT(IDAYY-1)+MNTT(IDAY
     2Y))/2.0
C
      IF(IT.LT.IRISE(IDAYY))TEMP(IT)=((MXTT(IDAYY-1)-MNTT(IDAYY))*(COS(P
     1I*(-(IT+10))/(10+IRISE(IDAYY)))))/2.0+(MXTT(IDAYY-1)+MNTT(IDAYY))/
     22.0
C
      IF(IT.EQ.IRISE(IDAYY))TEMP(IT)=MNTT(IDAYY)
C
      IF(IT.GT.IRISE(IDAYY).AND.IT.LT.14)TEMP(IT)=((MXTT(IDAYY)-MNTT(ID
     1AYY))*(-COS(PI*(IT-IRISE(IDAYY))/(14-IRISE(IDAYY)))))/2.0+(MXTT(I
     2DAYY)+MNTT(IDAYY))/2.0
C
      IF(IT.EQ.14)TEMP(IT)=MXTT(IDAYY)
C
      IF(IT.GT.14)TEMP(IT)=((MXTT(IDAYY)-MNTT(IDAYY+1))*(COS((PI*(14-IT
     1))/(10+IRISE(IDAYY+1)))))/2.0+(MXTT(IDAYY)+MNTT(IDAYY+1))/2.0
C
      IF(IT.EQ.1)AMTEMP(IT)=(TEMP(IT)+TEMP(24))/2.0
      IF(IT.GT.1)AMTEMP(IT)=(TEMP(IT)+TEMP(IT-1))/2.0
C
C ADJUSTMENTS ARE MADE FOR THE THRESHOLD TEMPERATURE BEING -3.6°C
C AND FOR HIGH AND LOW TEMPERATURES.
C
      HRDEG(IT)=AMTEMP(IT)+3.6
      IF(AMTEMP(IT).LE.10.0.AND.AMTEMP(IT).GT.5.0)HRDEG(IT)=
     1HRDEG(IT)*(AMTEMP(IT)-5.0)/5.0
      IF(AMTEMP(IT).LE.5.0)HRDEG(IT)=0.1
      IF(AMTEMP(IT).GT.22.5.AND.AMTEMP(IT).LE.25.0)HRDEG(IT)=26.1
      IF(AMTEMP(IT).GT.25.0.AND.AMTEMP(IT).LT.30.0)HRDEG(IT)=
     126.1*(30.0-AMTEMP(IT))/5.0
      IF(AMTEMP(IT).GE.30.0)HRDEG(IT)=0.1
C
 1320 CONTINUE
C
C********4.....IMMIGRATION******************
C
C THE BASIC DATA FOR THIS HAS ALREADY BEEN INPUT
C SKIP STATEMENTS
      IF(IDAYY.LT.IMSTAR)GO TO 26
      IF(IDAYY.GT.IMFINI)GO TO 26
      IF(IMM(IDAYY).EQ.0)GO TO 26
C
C THE NUMBER OF IMMIGRANTS/MILLION TILLERS IS CALCULATED, 64 IS THE
C RANDOM DEPOSITION FACTOR (TAYLOR AND PALMER 1972)
      ALATIM=IMM(IDAYY)*INCONF*TAYPAL*SEN6
C AND NOW PER TILLER
      ALTIM=ALATIM/1000000.0
      GO TO 27
   26 ALATIM=0.0
      ALTIM=0.0
   27 ALATAD(1)=ALATIM
      TOTALA=TOTALA+ALATAD(1)
C
C ********5.....DEVELOPMENT AND SURVIVAL*********
C
      DO 1000 IT=1,24
C
C SET LONGEVITIES , IN HOUR DEGREES, ARE INPUT FOR ALATE AND
C APTEROUS ADULTS AT DIFFERENT GROWTH STAGES
      SURTAL=2832.0*SEN3
```

```
          IF(GSTAGE.GT.59.0)SURTAL=2832.0*SEN3
          IF(GSTAGE.GT.73.0)SURTAL=2832.0*SEN3
          SURT=5664.0*SEN3
          IF(GSTAGE.GT.59.0)SURT=8496.0*SEN3
          IF(GSTAGE.GT.73.0)SURT=2832.0*SEN3
C NOW THE HOURLY SURVIVAL RATES, ALATES FIRST
          SARTA=10**(ALOG10(0.9*SEN4)/(SURTAL/(AMTEMP(IT)+3.6)))
          SURTA=10**(ALOG10(0.9*SEN4)/(SURT/(AMTEMP(IT)+3.6)))
          IF(GSTAGE.GT.73.0)SARTA=10**(ALOG10(0.6*SEN4)/(SURTAL/(AMTEMP
         1(IT)+3.6)))
          IF(GSTAGE.GT.73.0)SURTA=10**(ALOG10(0.6*SEN4)/(SURT/(AMTEMP(IT)
         1+3.6)))
C
C APTEROUS ADULTS ARE UPDATED. THE PROCESS IS DESCRIBED HERE IN DETAIL
C THE PROCEDURE FOR THE OTHER INSTARS AND MORPH IS THE SAME
C FIRST THE SKIP STATEMENT IF NO APHIDS PRESENT
          IF(TOTAD.EQ.0.0)GO TO 110
C THE TOTAL IS SET TO ZERO AS IS THE NUMBER OF REPRODUCING ADULTS
          TOTAD=0.0
          TOTADR=0.0
C NOW THE CHECK TO MAKE SURE THE LAST ELEMENT IN THE ARRAY IS EMPTY
          IF(.NOT.(ADULTS(750).EQ.0.0))GO TO 1001
C NOW THE UPDATING IS CARRIED OUT, STARTING WITH THE OLDEST APHIDS
          DO 109 J=1,749
          I=751-J
C ANOTHER SKIP STEP
          IF(ADULTS(I-1).EQ.0.0)GO TO 109
C THE NOS IN OLD AGE CLASS (I-1) ARE MOVED INTO (I) AND SOME DIE
          ADULTS(I)=ADULTS(I-1)*SURTA
C AGE IS UPDATED
          HRDAD(I)=HRDAD(I-1)+HRDEG(IT)
C THE TWO ARRAYS FOR (I-1) ARE ZEROED
          HRDAD(I-1)=0.0
          ADULTS(I-1)=0.0
C THG AGE FOR THAT ELEMENT IS CHECKED FOR LONGEVITY
          IF(HRDAD(I).LT.SURT)GO TO 210
C IF THE ADULTS ARE TOO OLD THEY DIE BUT WITH OTHER INSTARS THEY ARE
C MOVED INTO THE NEXT INSTAR 1ST AGE CLASS
          ADULTS(I)=0.0
          HRDAD(I)=0.0
C NOW THE ADULTS ARE TOTALLED
  210     TOTAD=TOTAD+ADULTS(I)
C NOW THE NUMBER OF REPRODUCING ADULTS
          IF(HRDAD(I).GT.(481.75*SEN3))TOTADR=TOTADR+ADULTS(I)
  109     CONTINUE
  110     CONTINUE
C
C NOW FOR THE ALATE ADULTS
C
          IF(TOTALA.EQ.0.0)GO TO 112
          TOTALA=0.0
          IF(.NOT.(ALATAD(750).EQ.0.0))GO TO 1001
          DO 111 J=1,749
          I=751-J
          IF(ALATAD(I-1).EQ.0.0)GO TO 111
          ALATAD(I)=ALATAD(I-1)*SARTA
          HRDALD(I)=HRDALD(I-1)+HRDEG(IT)
          HRDALD(I-1)=0.0
          ALATAD(I-1)=0.0
          IF(HRDALD(I).LT.SURTAL)GO TO 211
          ALATAD(I)=0.0
```

```
       HRDALD(I)=0.0
  211  TOTALA=TOTALA+ALATAD(I)
  111  CONTINUE
  112  CONTINUE

C
C NOW FOR THE 4TH INSTAR APTERAE
C
       SURN=10**(ALOG10(0.938*SEN4)/(4793.34*SEN3/(AMTEMP(IT)+3.6)))
       IF(GSTAGE.GT.73.0)SURN=10**(ALOG10(0.45*SEN4)/(4793.34*SEN3
      1/(AMTEMP(IT)+3.6)))
       IF(TOTFOR.EQ.0.0)GO TO 114
       TOTFOR=0.0
       ADULTS(1)=0.0
       IF(.NOT.(FNYMPH(200).EQ.0.0))GO TO 1001
       DO 113 J=1,199
       I=201-J
       IF(FNYMPH(I-1).EQ.0.0)GO TO 113
       FNYMPH(I)=FNYMPH(I-1)*SURN
       HRDFN(I)=HRDFN(I-1)+HRDEG(IT)
       FNYMPH(I-1)=0.0
       HRDFN(I-1)=0.0
       IF(HRDFN(I).LT.1300.61*SEN3)GO TO 212
       ADULTS(1)=ADULTS(1)+FNYMPH(I)
       FNYMPH(I)=0.0
       HRDFN(I)=0.0
       HRDAD(1)=0.0
  212  TOTFOR=TOTFOR+FNYMPH(I)
  113  CONTINUE
  114  CONTINUE
C
C NOW THE ALATE 4THS, THEY LIVE 1.5 X LONGER (HUGHES 1963).
C
       SURALN=10**(ALOG10(0.93*SEN4)/(5443.65*SEN3/(AMTEMP(IT)+3.6)))
       IF(GSTAGE.GT.73.0)SURALN=10**(ALOG10(0.374*SEN4)/(5443.65*SEN3
      1/(AMTEMP(IT)+3.6)))
       IF(TOTALF.EQ.0.0)GO TO 116
       TOTALF=0.0
       ALATAD(1)=0.0
       IF(.NOT.(ALFN(200).EQ.0.0))GO TO 1001
       DO 115 J=1,199
       I=201-J
       IF(ALFN(I-1).EQ.0.0)GO TO 115
       ALFN(I)=ALFN(I-1)*SURALN
       HRDALF(I)=HRDALF(I-1)+HRDEG(IT)
       ALFN(I-1)=0.0
       HRDALF(I-1)=0.0
       IF(HRDALF(I).LT.1950.92*SEN3)GO TO 213
       ALATED=ALATED+ALFN(I)
       ALFN(I)=0.0
       HRDALF(I)=0.0
       HRDALD(1)=0.0
  213  TOTALF=TOTALF+ALFN(I)
  115  CONTINUE
  116  CONTINUE
C
C PARASITISM AND DISEASE ARE NOW CONSIDERED
C
       IF(PARNO.EQ.1.0)GO TO 28
C THEY ARE CALCULATED DIRECTLY FROM FIELD DATA, PARASITISM BY DIVIDING
C THE PROPORTION KILLED BY 7.0 AND 24.0 TO GIVE APPROX HOURLY RATES,
```

```
C ONLY THE FIRST AGE CLASS OF THE ADULTS DIE
      TOTFAD=ALATED+ADULTS(1)
C THE PROCEDURE IS SKIPPED IF THIS ZERO
      IF(TOTFAD.EQ.0.0)GO TO 5432
      IF(IDAYY.LT.IPARA)GO TO 5432
      IF(IDAY.GT.IPAFIN)GO TO 5432
C NOW THE NUMBERS DYING ARE CALCULATED
      PARSIT=1000000.0*TOTDEN*PARA(IDAYY)*SEN8
      IF(PARSIT.GT.TOTFAD)PARSIT=TOTFAD
C NOW THE NUMBERS OF ALATES AND APTERAE KILLED ARE CALCULATED
      PARAL=ALATED*PARSIT/TOTFAD
      PARALD=ADULTS(1)*PARSIT/TOTFAD
      GO TO 5433
 5432 PARAL=0.0
      PARALD=0.0
 5433 ALATED=ALATED-PARAL
      ADULTS(1)=ADULTS(1)-PARALD
      IF(IT.EQ.13)TOTPAR=0.0
      TOTPAR=TOTPAR+((PARAL+PARALD)/1000000.0)
   28 CONTINUE
C
C ACCOUNT IS NOW MADE OF EMIGRATION
C
      IF(IT.EQ.13)TOTALE=0.0
      ALTEM=ALATED/1000000.0
      ALATEM=ALATED
      ALATED=0.0
      TOTALE=TOTALE+ALTEM
C
C NOW THE 3RD INSTAR APT
      IF(TOTTHI.EQ.0.0)GO TO 118
      TOTTHI=0.0
      FNYMPH(1)=0.0
      IF(.NOT.(TNYMPH(150).EQ.0.0))GO TO 1001
      DO 117 J=1,149
      I=151-J
      IF(TNYMPH(I-1).EQ.0.0)GO TO 117
      TNYMPH(I)=TNYMPH(I-1)*SURN
      HRDTN(I)=HRDTN(I-1)+HRDEG(IT)
      TNYMPH(I-1)=0.0
      HRDTN(I-1)=0.0
      IF(HRDTN(I).LT.1118.06*SEN3)GO TO 215
      FNYMPH(1)=FNYMPH(1)+TNYMPH(I)
      HRDTN(I)=0.0
      TNYMPH(I)=0.0
      HRDFN(1)=0.0
  215 TOTTHI=TOTTHI+TNYMPH(I)
  117 CONTINUE
  118 CONTINUE
C
C NOW THE ALATE 3RDS
      IF(TOTALT.EQ.0.0)GO TO 120
      TOTALT=0.0
      ALFN(1)=0.0
      IF(.NOT.(ALTN(150).EQ.0.0))GO TO 1001
      DO 119 J=1,149
      I=151-J
      IF(ALTN(I-1).EQ.0.0)GO TO 119
      ALTN(I)=ALTN(I-1)*SURALN
      HRDALT(I)=HRDALT(I-1)+HRDEG(IT)
      ALTN(I-1)=0.0
```

```
        HRDALT(I-1)=0.0
        IF(HRDALT(I).LT.1118.06*SEN3)GO TO 216
        ALFN(1)=ALFN(1)+ALTN(I)
        ALTN(I)=0.0
        HRDALT(I)=0.0
        HRDALF(1)=0.0
  216   TOTALT=TOTALT+ALTN(I)
  119   CONTINUE
  120   CONTINUE
C
C NOW THE 2ND INSTAR APTERAE
        IF(TOTSEC.EQ.0.0)GO TO 122
        TOTSEC=0.0
        TNYMPH(1)=0.0
        IF(.NOT.(SNYMPH(150).EQ.0.0))GO TO 1001
        DO 121 J=1,149
        I=151-J
        IF(SNYMPH(I-1).EQ.0.0)GO TO 121
        SNYMPH(I)=SNYMPH(I-1)*SURN
        HRDSN(I)=HRDSN(I-1)+HRDEG(IT)
        SNYMPH(I-1)=0.0
        HRDSN(I-1)=0.0
        IF(HRDSN(I).LT.1139.70*SEN3)GO TO 217
        TNYMPH(1)=TNYMPH(1)+SNYMPH(I)
        SNYMPH(I)=0.0
        HRDSN(I)=0.0
        HRDTN(1)=0.0
  217   TOTSEC=TOTSEC+SNYMPH(I)
  121   CONTINUE
  122   CONTINUE
C
C NOW THE ALATE 2NDS
        IF(TOTALS.EQ.0.0)GO TO 124
        TOTALS=0.0
        ALTN(1)=0.0
        IF(.NOT.(ALSN(150).EQ.0.0))GO TO 1001
        DO 123 J=1,149
        I=151-J
        IF(ALSN(I-1).EQ.0.0)GO TO 123
        ALSN(I)=ALSN(I-1)*SURALN
        HRDALS(I)=HRDALS(I-1)+HRDEG(IT)
        ALSN(I-1)=0.0
        HRDALS(I-1)=0.0
        IF(HRDALS(I).LT.1139.70*SEN3)GO TO 218
        ALTN(1)=ALTN(1)+ALSN(I)
        ALSN(I)=0.0
        HRDALS(I)=0.0
        HRDALT(1)=0.0
  218   TOTALS=TOTALS+ALSN(I)
  123   CONTINUE
  124   CONTINUE
C
C NOW THE 1ST INSTAR APTERAE
        IF(TOTFIR.EQ.0.0)GO TO 126
        TOTFIR=0.0
        SNYMPH(1)=0.0
        IF(.NOT.(PNYMPH(150).EQ.0.0))GO TO 1001
        DO 125 J=1,149
        I=151-J
        IF(PNYMPH(I-1).EQ.0.)GO TO 125
        PNYMPH(I)=PNYMPH(I-1)*SURN
```

81

```
        HRDPN(I)=HRDPN(I-1)+HRDEG(IT)
        PNYMPH(I-1)=0.0
        HRDPN(I-1)=0.0
        IF(HRDPN(I).LT.1234.97*SEN3)GO TO 219
        SNYMPH(1)=SNYMPH(1)+PNYMPH(I)
        PNYMPH(I)=0.0
        HRDPN(I)=0.0
        HRDSN(1)=0.0
  219   TOTFIR=TOTFIR+PNYMPH(I)
  125   CONTINUE
  126   CONTINUE
C
C NOW FINALLY THE ALATE 1STS
        IF(TOTALP.EQ.0.0)GO TO 128
        TOTALP=0.0
        ALSN(1)=0.0
        IF(.NOT.(ALPN(150).EQ.0.0))GO TO 1001
        DO 127 J=1,149
        I=151-J
        IF(ALPN(I-1).EQ.0.0)GO TO 127
        ALPN(I)=ALPN(I-1)*SURALN
        HRDALP(I)=HRDALP(I-1)+HRDEG(IT)
        ALPN(I-1)=0.0
        HRDALP(I-1)=0.0
        IF(HRDALP(I).LT.1234.97*SEN3)GO TO 220
        ALSN(1)=ALSN(1)+ALPN(I)
        ALPN(I)=0.0
        HRDALP(I)=0.0
        HRDALS(1)=0.0
  220   TOTALP=TOTALP+ALPN(I)
  127   CONTINUE
  128   CONTINUE
C TOTAL UP THE INSTARS
        TOTAD=TOTAD+ADULTS(1)
        TOTALA=TOTALA+ALATAD(1)
        TOTFOR=TOTFOR+FNYMPH(1)
        TOTALF=TOTALF+ALFN(1)
        TOTTHI=TOTTHI+TNYMPH(1)
        TOTALT=TOTALT+ALTN(1)
        TOTSEC=TOTSEC+SNYMPH(1)
        TOTALS=TOTALS+ALSN(1)
C**********6.....REPRODUCTION AND MORPH DETERMINATION****
C
C NOW REPRODUCTION/HD AT PARTICULAR DEV STAGES STARTING WITH THE
C APTERAE AND THEN THE ALATES. IT IS AFFECTED BY EXTREME TEMPS
        FEC=0.0062
        IF(GSTAGE.GT.59.0)FEC=0.0100
        IF(GSTAGE.GT.73.0)FEC=0.0062
        IF(GSTAGE.GT.83.0)FEC=0.0
        IF(AMTEMP(IT).LT.10.0)FEC=0.0
        IF(AMTEMP(IT).GT.20.0.AND.AMTEMP(IT).LT.30.0)FEC=FEC*((30.0-AMTEMP
       1(IT))/10.0)
        IF(AMTEMP(IT).GE.30.0)FEC=0.0
C NOW THE NYMPHS LAID BY THE APTERAE ARE CALCULATED
        NEWNY=TOTADR*FEC*(AMTEMP(IT)+3.6)*SEN5
C NOW THE ALATES
        ALFEC=0.0048
        IF(GSTAGE.GT.59.0)ALFEC=0.0079
        IF(GSTAGE.GT.73.0)ALFEC=0.0048
        IF(GSTAGE.GT.83.0)ALFEC=0.0
        IF(AMTEMP(IT).LT.10.0)ALFEC=0.0
```

82

```
      IF(AMTEMP(IT).GT.20.0.AND.AMTEMP(IT).LT.30.0)ALFEC=ALFEC*((30.0-
     1AMTEMP(IT))/10.0)
      IF(AMTEMP(IT).GE.30.0)ALFEC=0.0
C NOW THE NYMPHS LAID BY THE ALATES
      NWNY=TOTALA*ALFEC*(AMTEMP(IT)+3.6)*SEN5
C NOW ALL THE APHIDS ARE TOTALLED UP
      TOTDEN=(TOTAD+TOTALA+TOTFOR+TOTALF+TOTTHI+TOTALT+
     1TOTSEC+TOTALS+TOTFIR+TOTALP)/1000000.0
C NOW TO DECIDE THE PROPORTION OF ALATE NYMPHS
      ALATE=((2.603*TOTDEN+0.847208*GSTAGE-27.18896)/100.0)*SEN9
      IF(ALATE.GT.1.0)ALATE=1.0
      ALPN(1)=(NEWNY+NWNY)*ALATE
      IF(ALPN(1).LT.0.0)ALPN(1)=0.0
C NOW THE NUMBER OF APTERAE
      PNYMPH(1)=NEWNY+NWNY-ALPN(1)
      IF(PNYMPH(1).LT.0.0)PNYMPH(1)=0.0
C NOW THE TOTALS ARE CALCULATED
      TOTFIR=TOTFIR+PNYMPH(1)
      TOTALP=TOTALP+ALPN(1)
      HRDPN(1)=0.0
      HRDALP(1)=0.0
C NOW THE TOTALS ARE CONVERTED TO NOS PER TILLER
      PERADR=TOTADR/1000000.0
      PERAD=TOTAD/1000000.0
      PERALA=TOTALA/1000000.0
      PERF=TOTFOR/1000000.0
      PERALF=TOTALF/1000000.0
      PERT=TOTTHI/1000000.0
      PERALT=TOTALT/1000000.0
      PERS=TOTSEC/1000000.0
      PERALS=TOTALS/1000000.0
      PERP=TOTFIR/1000000.0
      PERALP=TOTALP/1000000.0
      TOTDEN=PERAD+PERALA+PERF+PERALF+PERT+PERALT+PERS+PERALS+
     1PERP+PERALP
C
C**********7.....PREDATORS*******************
C
C ALL INSTARS ARE CONVERTED INTO APHID UNITS
      AFIDUN=(PERAD+PERALA)+((PERF+PERALF)/1.5)+((PERT+PERALT)/2.0)
     1+((PERS+PERALS)/3.5)+((PERP+PERALP)/5.0)
      IF(PREDNO.EQ.1.0)GO TO 1223
      IF(IDAYY.LT.IPRED)GO TO 1222
      IF(IDAYY.GT.IPRFIN)GO TO 1222
C SKIP STEP FOR TEMPERATURE EXTREMES
      IF(AMTEMP(IT).LT.15.0)GO TO 716
      IF(AMTEMP(IT).GT.30.0)GO TO 716
      IF(AFIDUN.LE.0.0)GO TO 716
C NOW THE APHID UNITS IN THE FIRST THREE INSTARS
      ONYAFD=AFIDUN-((PERAD+PERALA)+(PERF+PERALF/1.5))
      IF(ONYAFD.LT.0.0)ONYAFD=0.0
      CALL PREDTR(AMTEMP(IT),PRED,IDAYY,AFIDUN,ONYAFD,PRDFAC,
     1PRDADC,TOTCON)
      PRDFAC=PRDFAC*SEN7
      PRDADC=PRDADC*SEN7
      IF(.NOT.(ALOWAF.EQ.1.0))GO TO 9999
      IF(ONYAFD.LE.3.0)PRDFAC=PRDFAC*3.0/ONYAFD
      ADUAFU=AFIDUN-ONYAFD
      IF(ADUAFU.LE.3.0)PRDADC=PRDADC*3.0/ADUAFU
 9999 IF(PRDFAC.GT.1.0)PRDFAC=1.0
      IF(PRDADC.GT.1.0)PRDADC=1.0
```

```
          GOTO 715
   716    PRDFAC=1.0
          PRDADC=1.0
   715    IF(PRDFAC.EQ.1.0.AND.PRDADC.EQ.1.0)GO TO 9998
C REDUCE NOS IN EACH INSTAR BECAUSE OF PREDATION
          DO 717 I=1,200
          FNYMPH(I)=FNYMPH(I)*PRDADC
          ALFN(I)=ALFN(I)*PRDADC
   717    CONTINUE
          DO 719 I=1,150
          TNYMPH(I)=TNYMPH(I)*PRDFAC
          ALTN(I)=ALTN(I)*PRDFAC
          SNYMPH(I)=SNYMPH(I)*PRDFAC
          ALSN(I)=ALSN(I)*PRDFAC
          PNYMPH(I)=PNYMPH(I)*PRDFAC
          ALPN(I)=ALPN(I)*PRDFAC
   719    CONTINUE
          DO 718 I=1,750
          ALATAD(I)=ALATAD(I)*PRDADC
          ADULTS(I)=ADULTS(I)*PRDADC
   718    CONTINUE
          PERAD=PERAD*PRDADC
          PERALA=PERALA*PRDADC
          PERF=PERF*PRDADC
          PERALF=PERALF*PRDADC
          PERT=PERT*PRDFAC
          PERALT=PERALT*PRDFAC
          PERS=PERS*PRDFAC
          PERALS=PERALS*PRDFAC
          PERP=PERP*PRDFAC
          PERALP=PERALP*PRDFAC
          TOTDEN=PERAD+PERALA+PERF+PERALF+PERT+PERALT+PERS+PERALS+
         1PERP+PERALP
  9998    CONTINUE
          IF(IT.EQ.13)DAYCON=0.0
          DAYCON=DAYCON+TOTCON
          GO TO 1223
  1222    TOTCON=0.0
          PRDFAC=1.0
          PRDADC=1.0
          DAYCON=0.0
  1223    CONTINUE
C
C***********8.......OUTPUT...***********************
C
          IF(.NOT.(IT.EQ.12))GO TO 135
          TOTYN=PERP+PERS+PERT+PERALP+PERALS+PERALT
          WRITE(2,132)IDAYY,PERP,PERS,PERT,PERF,PERAD,PERALP,PERALS,PERALT,P
         1ERALF,PERALA.TOTYN
   132    FORMAT(I4,11F10.4)
          WRITE(2,39)GSTAGE,PERADR,ALTIM,TOTALE,TOTDEN,TOTPAR,DAYCON,PRDFAC,
         1PRDADC,AFIDUN,TOT
    39    FORMAT(1 #10.4)
   135    CONTINUE
  1000    CONTINUE
C
C********9......CROP DEVELOPMENT MODEL...********
C
C THIS IS THE END OF THE DAY AND THE DEV. STAGE OF THE CROP IS UPDATED
          THRESH=6.0
          XMAX=MXTT(IDAYY)
```

84

```
      XMIN=MNTT(IDAYY)
      DD=0.0
      DO 8011 I=1,2
      Y=XMAX+XMIN-2.0*THRESH
      IF(XMIN.LT.THRESH)GO TO 6006
      B=0.25*Y
      GO TO 8010
 6006 IF(XMAX.GT.THRESH)GO TO 8008
      B=0.0
      GO TO 8010
 8008 T=ASIN(Y/(XMIN-XMAX))
      B=0.125*Y*(1.0-0.63661977*T)+0.079577472*(XMAX-XMIN)*COS(T)
      IF(B.LT.0.0)B=0.0
 8010 CONTINUE
      DD=DD+B
      XMIN=MNTT(IDAYY+1)
 8011 CONTINUE
C DAY DEGREES ARE SUMMED
      TOT=TOT+DD
      GSTAGE=0.173224*TOT-0.000125*TOT*TOT+26.33648
   91 CONTINUE
      IF(GGTAGE.GE.96.3)GO TO 1004
  107 CONTINUE
      GO TO 1003
C WARNING MESSAGE BECAUSE AN ARRAY IS OVERFLOING
 1001 WRITE(2,1002)
 1002 FORMAT(1H1, 15H ARRAY EXCEEDED)
 1003 CONTINUE
C
C*********...10...INPUT VARIABLES ARE PRINTED...*******
C
 1004 CONTINUE
      WRITE(2,1010)
 1010 FORMAT(35H CONC FACTORS AND SUCTION TRAP DATA////)
      WRITE(2,1012)INCONF,IMSTAR,IMFINI
 1012 FORMAT(3I4)
      WRITE(2,1013)(IMM(I),I=IMSTAR,IMFINI)
 1013 FORMAT(10I3)
      WRITE(2,1114)SEN1,SEN2,SEN3,SEN4,SEN5,SEN6,SEN7,SEN8,SEN9
 1114 FORMAT(9F5.2)
      WRITE(2,9997)PREDNO,PARNO,ALOWAF,LAT,TILERS,TAYPAL
 9997 FORMAT(6F10.2)
      IF(PREDNO.EQ.1.0)GO TO 7777
      WRITE(2,1014)
 1014 FORMAT(16H PREDATOR MATRIX///)
      WRITE(2,7780)IPRED,IPRFIN
 7780 FORMAT(2I4)
      DO 1016 I=IPRED,IPRFIN
      WRITE(2,1015)(PRED(I,J),J=1,5)
 1015 FORMAT(5F10.4)
 1016 CONTINUE
 7777 CONTINUE
      IF(PARNO.EQ.1.0)GO TO 7778
      WRITE(2,7781)IPARA,IPAFIN
 7781 FORMAT(2I4)
      WRITE(2,1017)
 1017 FORMAT(18H PARASITISM MATRIX///)
      WRITE(2,1018)(PARA(I),I=IPARA,IPAFIN)
 1018 FORMAT(5F10.4)
 7778 CONTINUE
      WRITE(2,9996)
```

```
 9996 FORMAT(10H MAX TEMPS//)
      WRITE(2,9995)(MXTT(I),I=ISTART-1,IFINIS)
 9995 FORMAT(20F7.2)
      WRITE(2,9994)
 9994 FORMAT(10H MIN TEMPS//)
      WRITE(2,9993)(MNTT(I),I=ISTART,IFINIS+1)
 9993 FORMAT(20F7.2)
C
C ********.....THE END....**********
C
      STOP
      END
      SUBROUTINE PREDIR(TEMP,PRED,I,AU,ONY,PRDFAC,PRDADC,TOTCON)
      DIMENSION PRED(250,5)
C PROP OF APHID UNITS IN INSTARS 1-3
      IF(ONY.EQ.0.0)GO TO 1111
      PERPRE=ONY/AU
C NO OF INSTARS 1-3 KILLED BY ALL COCCINELLID INSTARS
      TOTNY=(PRED(I,1)*0.0053+PRED(I,2)*0.0172+PRED(I,3)*0.0278
     1+(PRED(I,4)+PRED(I,5))*0.0477*PERPRE)*(TEMP+3.6)
C PROP OF TOTAL KILLED
      PRDFC=TOTNY/ONY
      IF(PRDFC.GT.1.0)PRDFC=1.0
C PROP SURVIVING
      PRDFAC=1.0-PRDFC
      GO TO 1112
C PROP OF INSTARS KILLED (4-5) BY COCC 4-5 IF NO OTHER PRED.
 1111 TOTADC=(PRED(I,4)+PRED(I,5))*0.0477*(TEMP+3.6)
      TOTNY=0.0
      PRDFAC=1.0
      GO TO 1113
C PROP OF INSTARS 4-5 PREYED ON IF THERE IS PRED OF 1-3
 1112 ADFAU=AU-ONY
      IF(ADFAU.LE.0.0)GO TO 4001
      TOTADC=(PRED(I,4)+PRED(I,5))*0.0477*(1.0-PERPRE)*(TEMP+3.6)
 1113 PRDAD=TOTADC/(AU-ONY)
      IF(PRDAD.GT.1.0)PRDAD=1.0
      PRDADC=1.0-PRDAD
      GO TO 4000
 4001 TOTADC=0.0
      PRDADC=0.0
 4000 CONTINUE
C SUM OF APHID UNITS CONSUMED
      TOTCON=TOTNY+TOTADC
      RETURN
      END
      FINISH
```

Appendix B

Definitions of variables used in model

Variable	Description	Unit
ADULTS(N)	Number of apterous aphids per million tillers	1
AFIDUN	Aphid units in number per tiller	1
ALATAD(N)	Number of alate adult aphids per million tillers	1
ALATE	Proportion of newly born aphid nymph that will develop into alates	1
ALATED	Number of recently moulted alate emigrant aphids per million tillers	1
ALATEM	Equivalent to ALATED	1
ALATIM	Immigration rate of alate aphids per million tillers	d^{-1}
ALFEC	Reproductive rate of alate adult aphids: nymphs alate^{-1} H°$^{-1}$	arbitrary unit
ALFN(N)	Number of fourth instar alatiform aphid nymphs per million tillers	1
ALOWAF	Switch variable for inclusion exclusion of aphid density threshold with predation	—
ALPN(N)	Number of first-instar alatiform aphid nymphs per million tillers	1
ALSN(N)	Number of second-instar alatiform aphid nymphs per million tillers	1
ALTEM	Number of emigrant alate aphids per tiller	1
ALTIM	Number of immigrant alate aphids per tiller	1
ALTN(N)	Number of third-instar alatiform aphid nymphs per million tillers	1
AMTEMP(N)	Mean temperatures (time interval 1h)	°C
DAYCON	Consumption rate of coccinellids in aphid units	d^{-1}
DAYL(N)	Day length	h
DD	Daily physiological time units for crop development: D°	arbitrary unit
FEC	Reproductive rate of apterous adult aphids in nymphs aptera^{-1} H°$^{-1}$	arbitrary unit
FNYMPH(N)	Number of fourth-instar apteriform aphid nymphs per million tillers	1
GSTAGE	Crop developmental stage (decimal code)	arbitrary unit
HRDAD(N)	Physiological age of apterous adult aphids: H°	,,
HRDALD(N)	Physiological age of alate adult aphids: H°	,,
HRDALF(N)	Physiological age of fourth-instar alatiform aphid nymphs: H°	,,
HRDALP(N)	Physiological age of first-instar alatiform aphid nymphs: H°	,,

Variable	Description	Unit
HRDALS(N)	Physiological age of second-instar alatiform aphid nymphs: H°	,,
HRDALT(N)	Physiological age of third-instar alatiform aphid nymphs: H°	,,
HRDEG(N)	Physiological time: H°, each hour	,,
HRDFN(N)	Physiological age of fourth-instar apteriform aphid nymphs: H°	,,
HRDPN (N)	Physiological age of first-instar apteriform aphid nymphs: H°	,,
HRDSN(N)	Physiological age of second-instar apteriform aphid nymphs: H°	,,
HRDTN(N)	Physiological age of third-instar apteriform aphid nymphs: H°	,,
IDAYY	Day number: 1 January = Day 1	—
IFINIS	Day number at end of simulation	—
IMFINI	Day number at end of aphid immigration	—
IMM(N)	Number of alate adult aphids caught in a 12.2 m suction trap	d^{-1}
IMSTAR	Day number at start of aphid immigration	—
INCONF	Concentration factor for aphids settling in a crop in excess of random settling	—
IPARA	Day number at start of parasitism and disease	—
IPAFIN	Day number at end of parasitism and disease	—
IPRED	Day number at start of predation	—
IPRFIN	Day number at end of predation	—
IRISE(N)	Time of sunrise, as an integer	—
ISTART	Day number at start of simulation	—
LAT	Latitude of site	degrees of angle
MNTT(N)	Minimum temperature (time inverval 1 d)	°C
MXTT(N)	Maximum temperature (time interval 1 d)	°C
NEWNY	Number of newly born aphid nymphs with apterous mothers per million tillers	1
NWNY	Number of newly born aphid nymphs with alate mothers per million tillers	1
ONYAFD	Number of aphid units in first three aphid instars per million tillers	1
PARA(N)	Number of aphids dying from parasitism and disease per tiller, constant over the day	h^{-1}
PARAL	Number of newly moulted alate adult aphids dying from parasitism and disease per million tillers	h^{-1}
PARALD	Number of newly moulted apterous adult aphids dying from parasitism and disease per million tillers	h^{-1}
PARNO	Switch variable for inclusion or exclusion of mortality due to parasitism and disease	—
PARSIT	Number of newly moulted adult aphids dying from parasi-	

Variable	Description	Unit
	tism and disease per million tillers	h^{-1}
PERAD	Number of apterous adult aphids per tiller	1
PERADR	Number of reproductively mature apterous adult aphids per tiller	1
PERALA	Number of alate adult aphids per tiller	1
PERALF	Number of fourth-instar alatiform aphid nymphs per tiller	1
PERALP	Number of first-instar alatiform aphid nymphs per tiller	1
PERALS	Number of second-instar alatiform aphid nymphs per tiller	1
PERALT	Number of third-instar alatiform aphid nymphs per tiller	1
PERF	Number of fourth-instar apteriform aphid nymphs per tiller	1
PERP	Number of first-instar apteriform aphid nymphs per tiller	1
PERS	Number of second-instar apteriform aphid nymphs per tiller	1
PERT	Number of third-instar apteriform aphid nymphs per tiller	1
PNYMPH(N)	Number of first-instar apteriform aphid nymphs per million tillers	1
PRDADC	Proportion of fourth-instar and adult aphids surviving predation	h^{-1}
PRDFAC	Proportion of the first three aphid instars surviving predation	h^{-1}
PRED(N_1, N_2)	Number of coccinellids, in individual instars per tiller	1
PREDNO	Switch variable for inclusion or exclusion of mortality due to coccinellids	—
RISE(N)	Time of sunrise	—
SARTA	Survival rate of alate adult aphids	h^{-1}
SNYMPH(N)	Density of second-instar apteriform aphid nymphs per million tillers	1
SURALN	Survival rate of alatiform aphid nymphs	h^{-1}
SURN	Survival rate of apteriform aphid nymphs	h^{-1}
SURT	Longevity of apterous adult aphids	arbitrary unit
SURTA	Survival rate of apterous adult aphids	h^{-1}
SURTAL	Longevity of alate adult aphids	arbitrary unit
TAYPAL	Taylor-Palmer random deposition factor: number of alate adult aphids settling at random per million tillers per alate in a suction trap	d^{-1}
TEMP(N)	Temperature each hour	°C
THRESH	Crop development threshold temperature	°C
TILERS	Number of tillers	m^{-2}
TNYMPH(N)	Number of third-instar apteriform aphid nymphs per million tillers	1
TOT	Accumulated physiological time units for crop development	arbitrary unit
TOTAD	Number of apterous adult aphids per million tillers	1
TOTADR	Number of reproductively mature apterous adult aphids per million tillers	1

Variable	Description	Unit
TOTALA	Number of alate adult aphids per million tillers	1
TOTALE	Number of alate adult emigrant aphids per tiller	1
TOTALF	Number of fourth-instar alatiform aphid nymphs per million tillers	1
TOTALP	Number of first-instar alatiform aphid nymphs per million tillers	1
TOTALS	Number of second-instar alatiform aphid nymphs per million tillers	1
TOTALT	Number of third-instar alatiform aphid nymphs per million tillers	1
TOTCON	Number of aphid units consumed by coccinellids	h^{-1}
TOTDEN	Number of aphids per tiller	1
TOTFAD	Number of newly moulted adult aphids per million tillers	1
TOTFIR	Number of first-instar apteriform aphid nymphs per million tillers	1
TOTFOR	Number of fourth-instar apteriform aphid nymphs per million tillers	1
TOTPAR	Calculated number of newly moulted adult aphids dying from parasitism and disease per tiller	d^{-1}
TOTSEC	Number of second-instar apteriform aphid nymphs per million tillers	1
TOTTHI	Number of third-instar apteriform aphid nymphs per million tillers	1
TOTYN	Number of aphids in the first three instars per tiller	1

Predator subroutine

Variable	Description	Unit
ADFAU	Number of aphid units in fourth-instar and adult aphids per tiller	1
AU	Number of aphid units per tiller	1
ONY	Number of aphid units in first three aphid instars per tiller	1
PERPRE	Proportion of aphid units in first three aphid instars	—
PRDAD	Proportion of aphid units in fourth instar and adult aphids dying from predation	h^{-1}
PRDADC	Proportion of aphid units in fourth instar and adult aphids surviving predation	h^{-1}
PRDFAC	Proportion of aphid units in first three aphid instars surviving predation	h^{-1}
PRDFC	Proportion of aphid units in first three aphid instars dying from predation	h^{-1}
PRED(N_1,N_2)	Number of coccinellids, in individual instars per tiller	1
TEMP	Mean hourly temperature	°C
TOTADC	Number of aphid units in fourth instar and adult aphids consumed by coccinellids per tiller	h^{-1}

Variable	Description	Unit
TOTNY	Number of aphid units in first three instars consumed by coccinellids per tiller	h^{-1}